Yamaha 650 Twins Owners Workshop Manual

by Pete Shoemark
with an additional Chapter on 1977 to 1983 models
by Penny Cox

Models covered
XS1, XS1B and XS2. 653cc. US 1970 to 1973
TX650. 653cc. US 1973 to 1974
XS650 B, C and D. 653cc. US 1974 to 1977
XS650 E and SE. 653cc. US 1978
XS650 F, SF and 2F. 653cc. US 1979
XS650 G and SG. 653cc. US 1980
XS650 H and SH. 653cc. US 1981
XS650 SJ. 653cc. US 1982
XS650 SK. 653cc. US 1983
XS650. 653cc. UK 1975 to 1981
XS650 SE Special/US Custom. 653cc. UK 1979 to 1982

ISBN 978 1 85010 921 1

(341-5S7)

THE BOOK

J H Haynes & Co. Ltd.
Haynes North America, Inc

www.haynes.com

Acknowledgements

Our thanks are due to Jim Patch of Yeovil Motor Cycle Services for his help in supplying technical information and allowing us to take the photographs included in the update Chapter of this manual, to R. S. Damerell and Son Ltd of St. Austell, Cornwall, who supplied the machine featured in the main text, and to Mr Lee Taft of Yamaha of Thousand Oaks, Thousand Oaks, California who provided the machine for the cover photograph.

Thanks are also due to the Avon Rubber Company, who kindly supplied information and technical assistance on tyre fitting; NGK Spark Plugs (UK) Limited for information on sparking plug maintenance and electrode conditions and to Renold Limited for advice on chain care and renewal.

Finally, thanks must go to all those people at Sparkford who assisted in the production of this manual, particularly Brian Horsfall who assisted with the stripdown and rebuilding and devised the ingenious methods for overcoming the lack of service tools; Les Brazier and Tony Steadman who arranged and took the photographs, and Mansur Darlington who edited the text.

About this manual

The author of this manual has the conviction that the only way in which a meaningful and easy to follow text can be written is first to do the work himself, under conditions similar to those found in the average household. As a result, the hands seen in the photographs are those of the author. Unless specially mentioned, and therefore considered essential, Yamaha service tools have not been used. There is invariably some alternative means of loosening or removing a vital component when service tools are not available, but risk of damage should always be avoided.

Each of the six Chapters is divided into numbered sections. Within these sections are numbered paragraphs. Cross reference throughout the manual is quite straightforward and logical. When reference is made 'See Section 6.10' it means Section 6, paragraph 10 in the same Chapter. If another Chapter were intended, the reference would read, for example, 'See Chapter 2, Section 6.10'. All the photographs are captioned with a section/ paragraph number to which they refer and are relevant to the Chapter text adjacent.

Figures (usually line illustrations) appear in a logical but numerical order, within a given Chapter. Fig. 1.1 therefore refers to the first figure in Chapter 1.

Left-hand and right-hand descriptions of the machines and their components refer to the left and right of a given machine when the rider is seated normally.

Motorcycle manufactures continually make changes to specifications and recommendations, and these, when notified, are incorporated into our manuals at the earliest opportunity.

We take great pride in the accuracy of information given in this manual, but motorcycle manufacturers make alterations and design changes during the production run of a particular motorcycle of which they do not inform us. No liability can be accepted by the authors or publishers for loss, damage or injury caused by any errors in, or omissions from, the information given.

Introduction to the Yamaha 650 twins

Although the history of Yamaha can be traced back to the year 1887, when a then very small company commenced manufacture of reed organs, it was not until 1954 that the company became interested in motor cycles. As can be imagined, the problems of marketing a motor cycle against a background of musical instruments manufacture were considerable. Some local racing successes and the use of hitherto unknown bright colour schemes helped achieve the desired results and in July 1955 the Yamaha Motor Company was established as a separate entity, employing a work force of less than 100 and turning out some 300 machines a month.

Competition successes continued and with the advent of tasteful styling that followed Italian trends, Yamaha became established as one of the world's leading motor cycle manufacturers. Part of this success story is the impressive list of Yamaha 'firsts' - a whole string of innovations that include electric starting, pressed steel frame, torque induction and 6 and 8 port engines. There is also the "Autolube" system of lubrication, in which the engine-driven oil pump is linked to the twist grip throttle, so that lubrication requirements are always in step with engine demands.

Since 1964, Yamaha has gained the World Championship on numerous occasions, in both the 125 cc and 250 cc classes. Indeed, Yamaha has dominated the lightweight classes in international road racing events to such an extent in recent years that several race promoters are now instituting a special type of event in their programme from which Yamaha machines are barred! Most of the racing successes have been achieved with twin cylinder two-strokes and the practical experience gained has been applied to the road going versions.

In recent years, Yamaha have begun to make inroads into the four-stroke sector of the market, the 650 twin being one of the longest running models of this type. The original XS 1 model was first introduced in 1970. At that time it was equipped with a twin leading shoe front brake, but other than this and the many minor improvements over the years, it remains basically unchanged to date.

The 645 cc single overhead camshaft engine has been changed only in detail, to obtain better performance and smoother running, the basic vertical twin configuration having been retained.

Contents

Left-hand view of 1977 Yamaha XS 650

Ordering spare parts

When ordering spare parts for the Yamaha 650 cc twins, it is advisable to deal direct with an official Yamaha agent, who will be able to supply many of the items required ex-stock. Although parts can be ordered from Yamaha direct, it is preferable to route the order via a local agent even if the parts are not available from stock. He is in a better position to specify exactly the parts required and to identify the relevant spare part numbers so that there is less chance of the wrong part being supplied by the manufacturer due to a vague or incomplete description.

When ordering spares, always quote the frame and engine numbers in full, together with any prefixes or suffixes in the form of letters. The frame number is found stamped on the right-hand side of the steering head, in line with the forks. The engine number is stamped on the front of the crankcase.

Use only parts of genuine Yamaha manufacture. A few pattern parts are available, sometimes at cheaper prices, but there is no guarantee that they will give such good service as the originals they replace. Retain any worn or broken parts until the replacements have been obtained; they are sometimes needed as a pattern to help identify the correct replacement when design changes have been made during a production run.

Some of the more expendable parts such as spark plugs, bulbs, tyres, oils and greases etc., can be obtained from accessory shops and motor factors, who have convenient opening hours and can often be found not far from home. It is also possible to obtain parts on a Mail Order basis from a number of specialists who advertise regularly in the motor cycle magazines.

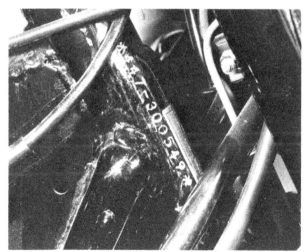

Location of Frame No

Location of Engine No

Safety first!

Professional motor mechanics are trained in safe working procedures. However enthusiastic you may be about getting on with the job in hand, do take the time to ensure that your safety is not put at risk. A moment's lack of attention can result in an accident, as can failure to observe certain elementary precautions.

There will always be new ways of having accidents, and the following points do not pretend to be a comprehensive list of all dangers; they are intended rather to make you aware of the risks and to encourage a safety-conscious approach to all work you carry out on your vehicle.

Essential DOs and DON'Ts

DON'T start the engine without first ascertaining that the transmission is in neutral.

DON'T suddenly remove the filler cap from a hot cooling system – cover it with a cloth and release the pressure gradually first, or you may get scalded by escaping coolant.

DON'T attempt to drain oil until you are sure it has cooled sufficiently to avoid scalding you.

DON'T grasp any part of the engine, exhaust or silencer without first ascertaining that it is sufficiently cool to avoid burning you.

DON'T allow brake fluid or antifreeze to contact the machine's paintwork or plastic components.

DON'T syphon toxic liquids such as fuel, brake fluid or antifreeze by mouth, or allow them to remain on your skin.

DON'T inhale dust – it may be injurious to health (see *Asbestos* heading).

DON'T allow any spilt oil or grease to remain on the floor – wipe it up straight away, before someone slips on it.

DON'T use ill-fitting spanners or other tools which may slip and cause injury.

DON'T attempt to lift a heavy component which may be beyond your capability – get assistance.

DON'T rush to finish a job, or take unverified short cuts.

DON'T allow children or animals in or around an unattended vehicle.

DON'T inflate a tyre to a pressure above the recommended maximum. Apart from overstressing the carcase and wheel rim, in extreme cases the tyre may blow off forcibly.

DO ensure that the machine is supported securely at all times. This is especially important when the machine is blocked up to aid wheel or fork removal.

DO take care when attempting to slacken a stubborn nut or bolt. It is generally better to pull on a spanner, rather than push, so that if slippage occurs you fall away from the machine rather than on to it.

DO wear eye protection when using power tools such as drill, sander, bench grinder etc.

DO use a barrier cream on your hands prior to undertaking dirty jobs – it will protect your skin from infection as well as making the dirt easier to remove afterwards; but make sure your hands aren't left slippery. Note that long-term contact with used engine oil can be a health hazard.

DO keep loose clothing (cuffs, tie etc) and long hair well out of the way of moving mechanical parts.

DO remove rings, wristwatch etc, before working on the vehicle – especially the electrical system.

DO keep your work area tidy – it is only too easy to fall over articles left lying around.

DO exercise caution when compressing springs for removal or installation. Ensure that the tension is applied and released in a controlled manner, using suitable tools which preclude the possibility of the spring escaping violently.

DO ensure that any lifting tackle used has a safe working load rating adequate for the job.

DO get someone to check periodically that all is well, when working alone on the vehicle.

DO carry out work in a logical sequence and check that everything is correctly assembled and tightened afterwards.

DO remember that your vehicle's safety affects that of yourself and others. If in doubt on any point, get specialist advice.

IF, in spite of following these precautions, you are unfortunate enough to injure yourself, seek medical attention as soon as possible.

Asbestos

Certain friction, insulating, sealing, and other products – such as brake linings, clutch linings, gaskets, etc – contain asbestos. *Extreme care must be taken to avoid inhalation of dust from such products since it is hazardous to health.* If in doubt, assume that they *do* contain asbestos.

Fire

Remember at all times that petrol (gasoline) is highly flammable. Never smoke, or have any kind of naked flame around, when working on the vehicle. But the risk does not end there – a spark caused by an electrical short-circuit, by two metal surfaces contacting each other, by careless use of tools, or even by static electricity built up in your body under certain conditions, can ignite petrol vapour, which in a confined space is highly explosive.

Always disconnect the battery earth (ground) terminal before working on any part of the fuel or electrical system, and never risk spilling fuel on to a hot engine or exhaust.

It is recommended that a fire extinguisher of a type suitable for fuel and electrical fires is kept handy in the garage or workplace at all times. Never try to extinguish a fuel or electrical fire with water.

Note: *Any reference to a 'torch' appearing in this manual should always be taken to mean a hand-held battery-operated electric lamp or flashlight. It does **not** mean a welding/gas torch or blowlamp.*

Fumes

Certain fumes are highly toxic and can quickly cause unconsciousness and even death if inhaled to any extent. Petrol (gasoline) vapour comes into this category, as do the vapours from certain solvents such as trichloroethylene. Any draining or pouring of such volatile fluids should be done in a well ventilated area.

When using cleaning fluids and solvents, read the instructions carefully. Never use materials from unmarked containers – they may give off poisonous vapours.

Never run the engine of a motor vehicle in an enclosed space such as a garage. Exhaust fumes contain carbon monoxide which is extremely poisonous; if you need to run the engine, always do so in the open air or at least have the rear of the vehicle outside the workplace.

The battery

Never cause a spark, or allow a naked light, near the vehicle's battery. It will normally be giving off a certain amount of hydrogen gas, which is highly explosive.

Always disconnect the battery earth (ground) terminal before working on the fuel or electrical systems.

If possible, loosen the filler plugs or cover when charging the battery from an external source. Do not charge at an excessive rate or the battery may burst.

Take care when topping up and when carrying the battery. The acid electrolyte, even when diluted, is very corrosive and should not be allowed to contact the eyes or skin.

If you ever need to prepare electrolyte yourself, always add the acid slowly to the water, and never the other way round. Protect against splashes by wearing rubber gloves and goggles.

Mains electricity and electrical equipment

When using an electric power tool, inspection light etc, always ensure that the appliance is correctly connected to its plug and that, where necessary, it is properly earthed (grounded). Do not use such appliances in damp conditions and, again, beware of creating a spark or applying excessive heat in the vicinity of fuel or fuel vapour. Also ensure that the appliances meet the relevant national safety standards.

Ignition HT voltage

A severe electric shock can result from touching certain parts of the ignition system, such as the HT leads, when the engine is running or being cranked, particularly if components are damp or the insulation is defective. Where an electronic ignition system is fitted, the HT voltage is much higher and could prove fatal.

Routine maintenance

Refer to Chapter 7 for details of 1977 to 1983 models

Introduction

Periodic routine maintenance is a continuous process that commences immediately the machine is used. It must be carried out at specified mileage recording, or on a calendar basis if the machine is not used frequently, whichever is the sooner. Maintenance should be regarded as an insurance policy, to help keep the machine in the peak of condition and to ensure long, trouble-free service. It has the additional benefit of giving early warning of any faults that may develop and will act as a regular safety check, to the obvious advantage of both rider and machine alike.

The various maintenance tasks are described under their respective mileage and calendar headings. Accompanying diagrams are provided, where necessary. It should be remembered that the interval between the various maintenance tasks serves only as a guide. As the machine gets older or is used under particularly adverse conditions, it would be advisable to reduce the period between each check.

For ease of reference each service operation is described in detail under the relevant heading. However, if further general information is required, it can be found within the manual under the pertinent section heading in the relevant Chapter.

In order that the routine maintenance tasks are carried out with as much ease as possible, it is essential that a good selection of general workshop tools are available.

Included in the kit must be a range of metric ring or combination spanners, a selection of crosshead screwdrivers and at least one pair of circlip pliers.

Additionally, owing to the extreme tightness of most casing screws on Japanese machines, an impact screwdriver, together with a choice of large or small crosshead screw bits, is absolutely indispensable. This is particularly so if the engine has not been dismantled since leaving the factory.

Weekly or every 250 miles

1   Topping up engine oil

Remove the combined oil filler cap/dipstick from the crankcase, and wipe the dipstick clean. Place the dipstick back in position, but do not not screw it home. Remove and check the oil level. On early models, the dipstick has three marks indicating the maximum, recommended and minimum oil levels. Later models have only two marks, indicating maximum and minimum levels. Top up as necessary, using SAE 20W/50 engine oil. Note that on early models, any tendency to produce excessive oil mist from the breather pipes may be rectified by lowering the recommended oil level mark (on early type dipstick only) by 10 mm.

Check oil level with dipstick

2   Tyre pressures

Check the tyre pressures with a pressure gauge that is known to be accurate. Always check the pressure when the tyres are cold. If the machine has travelled a number of miles, the tyres will have become hot and consequently the pressure will have increased. A false reading will therefore result.

Tyre pressures

Front	Rear
23 psi (1.6 kg/cm^2)	28 psi (2.0 kgs/cm^2)

3   Hydraulic fluid level - disc brake models

Check the level of the fluid in the master cylinder reservoir. On machines fitted with a tamper-proof cover, this may be observed through the translucent side of the reservoir. On models fitted with a screw cap, remove the cap to observe the fluid level. During normal service, it is unlikely that the hydraulic fluid level will fall dramatically, unless a leak has developed in the system. If this occurs, the fault should be remedied **at once.** The level will fall slowly as the brake linings wear and the fluid deficiency should be corrected, when required. Always use an hydraulic fluid of DOT 3 or SAE J1703 specification, and do not mix different types of fluid, even if the specifications appear the same. This will preclude the possibility of two incompatible fluids being mixed and the resultant chemical reaction damaging the seals.

If the level in the reservoir has been allowed to fall below the specified limit, and air has entered the system, the brake in question must be bled, as described in Chapter 5.

4 *Safety check*

Give the machine a close visual inspection, checking for loose nuts and fittings, frayed control cables etc.

5 *Legal check*

Ensure that the lights, horn and traffic indicators function correctly, also the speedometer.

Check and top up hydraulic fluid level

Monthly or every 500 miles

Complete the tasks listed under the weekly/250 mile heading and then carry out the following checks.

1 *Tyre damage*

Rotate each wheel and check for damage to the tyres, especially splitting on the sidewalls. Remove any stones or other objects caught between the treads. This is particularly important on the front tyre, where rapid tyre deflation due to penetration of the inner tube will almost certainly cause total loss of control of the machine.

2 *Spoke tension*

Check the spokes for tension, by gently tapping each one with a metal object. A loose spoke is identifiable by the low pitch noise emitted when struck. If any one spoke needs considerable tightening, it will be necessary to remove the tyre and inner tube in order to file down the protruding spoke end. This will prevent it from chafing through the rim band and piercing the inner tube.

3 *Front brake adjustment (drum brake models)*

Check the front brake for free play at the handlebar lever and for efficient operation. Adjustment for excess free play should be carried out as follows:

Place the machine on the centre stand so that the front wheel is raised clear of the ground. Remove the clevis pin from one end of the rod which connects the two actuating arms. Have an assistant operate the handlebar lever so that one shoe is *just* in contact with the drum surface, then operate the second actuating lever by hand, so that the second shoe is in light contact. The hole in the clevis should now align with that of the actuating arm. If this is not the case, adjust the length of the connecting rod until this setting is acheived. It is important that both shoes commence operation at the same point if full braking efficiency is to be expected. Finally, set the free play in the operating cable to give 6 mm (¼ in) slack.

4 *Rear brake adjustment*

When the rear brake is in correct adjustment the total brake pedal travel measured at the toe tread should be about 25 mm (1 in). If the travel is greater or less than this carry out the necessary adjustment by means of the shouldered nut at the brake arm end of the cable.

5 *Final drive chain lubrication*

In order that final drive chain life is extended as much as possible, regular lubrication and adjustment is essential. This is particularly so when the chain is not enclosed or is fitted to a machine transmitting high power to the rear wheel. The chain may be lubricated whilst it is in place on the machine by the application of one of the proprietary chain greases contained in an aerosol can. Ordinary engine oil can be used, though owing to the speed with which it is flung off the rotating chain, its effective life is limited.

The most satisfactory method of chain lubrication can be made when the chain has been removed from the machine. Clean the chain in paraffin and wipe it dry. The chain can now be immersed in one of the special chain graphited greases. The grease must be heated as per the instructions on the can so that the lubricant penetrates into the areas between the link pins and the rollers.

6 *Final drive chain adjustment*

Check the slack in the final drive chain. The correct up and down movement, as measured at the mid-point of the chain lower run, should be ¾ in (20 mm). Adjustment should be carried out as follows: Place the machine on the centre stand so that the rear wheel is clear of the ground and free to rotate. Remove the split pin from the wheel spindle and slacken the wheel nut a few turns. Loosen the locknuts on the two chain adjuster bolts. Rotation of the adjuster bolts in a clockwise direction will tighten the chain. Tighten each bolt a similar number of turns so that wheel alignment is maintained. This can be verified by checking that the mark on the outer face of each chain adjuster is aligned with the same aligning mark on each fork end. With the adjustment correct, tighten the wheel nut and fit a new split pin. Finally, retighten the adjuster bolt locknuts.

Use aerosol chain lubricant frequently

Three monthly or every 2000 miles

Complete the checks listed under the weekly/250 mile and monthly/500 mile headings and then carry out the following tasks:

1 Air filter cleaning

Remove the side panels to expose the air cleaner cases. Release the side flaps to gain access to the filter elements.

Where paper element, or dry foam type filters are employed, tap the element to loosen the accumulated dust, then blow the element from the inside, using compressed air.

Where wet foam elements are used, wash them thoroughly in detergent and water. Dry the elements thoroughly, and then soak them in fresh engine oil, squeezing them to remove any excess. Installation is a reversal of the removal sequence.

2 Carburettor synchronisation and setting

Check and, if necessary adjust, the carburettor settings and synchronisation. This is covered in detail in Chapter 2 of this manual.

3 Cleaning and adjusting the contact breaker points

Remove the contact breaker inspection cover and gasket. The cover is retained by two screws. Inspect the faces of the two sets of contact breaker points. Slight pitting or burning can be removed while the contact breaker unit is in place on the machine, using a very fine swiss file or emery paper (No 400) backed by a thin strip of tin. If the pitting or burning is excessive, the contact breaker unit in question should be removed for points dressing or renewal (see Chapter 3, Section 4).

Rotate the engine until one set of points is in the fully open position. The correct gap is within the range 0.3 - 0.4 mm (0.012 - 0.016 in). Adjustment is effected by slackening the screw holding the fixed contact breaker point in position and moving the point either closer to or further away with a screwdriver inserted between the small upright post and the slot in the fixed contact plate. Make sure that the points are in the fully open position when this adjustment is made or a false reading will result. When the gap is correct, tighten the screw and recheck.

Repeat the procedure with the other set of points.

4 Checking the ignition timing

While the contact breaker cover is removed, the ignition timing setting should be checked. Remove the alternator cover, and turn the engine slowly until the scribe line on the alternator rotor is in line with the F mark on the stator. At this point the contact breakers for one cylinder should just be separating. Rotate the engine through 360° and check the other cylinder. If the setting appears incorrect, reset the timing as described in Chapter 3 Section 7.

5 Final drive chain lubrication

The final drive chain should be removed from the machine for thorough cleaning and lubrication if long service life is to be expected. This is in addition to the intermediate lubrication carried out with the chain on the machine, as described under the weekly/250 mile servicing heading.

Separate the chain by removing the master link and run it off the sprockets. If an old chain is available, interconnect the old and new chain, before the new chain is run off the sprockets. In this way the old chain can be pulled into place on the sprockets and then used to pull the regreased chain into place with ease.

Clean the chain thoroughly in a paraffin bath and then finally with petrol. The petrol will wash the paraffin out of the links and rollers which will then dry more quickly.

Allow the chain to dry and then immerse it in a molten lubricant such as Linklyfe or Chainguard. These lubricants must be used hot to achieve better penetration of the links and rollers. They are less likely to be thrown off by centrifugal force when the chain is in motion.

Refit the newly greased chain onto the sprocket, replacing the master link. This is accomplished most easily when the free ends of the chain are pushed into mesh on the rear wheel sprocket. The spring link must be fitted so that the closed end faces the direction of chain travel.

Filter elements pull out for cleaning

Check contact breaker gap using a feeler gauge

Points should **just** start to separate at this position

6 General lubrication

Apply grease or oil to the handlebar lever pivots and to the centre stand and prop stand pivots.

7 Control cable lubrication

Lubricate the control cables thoroughly with motor oil or an all-purpose oil. A good method of lubricating the cables is shown in the accompanying illustration.

8 Changing the engine oil

Obtain a container of at least three litres capacity into which the engine oil can be drained. Warm the engine up thoroughly, then remove the two crankcase drain plugs and leave the oil to drain thoroughly. It is recommended that the two gauze filters are removed and cleaned at each oil change(See Chapter 2 Section 12).

Finally, refit the drain plugs and refill to the recommended level with SAE 20W/50 engine oil. The oil capacities are as follows:

XS1B	3.2 litres	5.6 pints (3.4 US qts)
XS2 TX650	2.6 litres	4.6 pints (2.7 US qts)
TX650A XS650B XS650C	2.5 litres	4.4 pints (2.6 US qts)

9 Swinging arm pivot shaft lubrication

Using a grease gun, inject high melting point grease into the grease nipple at each end of the pivot shaft, until it just starts to exude around the seals.

10 Decarbonising the engine

Remove the cylinder head and carefully scrape off all carbon deposits on the piston crowns and in the combustion chambers. If necessary, remove and regrind the valves. Refer to Chapter One for details.

11 Valve clearance adjustment

Regardless of whether or not the cylinder head has been removed for decarbonisation, the valve clearances should be checked, and if necessary, adjusted.

Remove the four inspection covers from the cylinder head cover, and also the alternator outer cover. Turn the engine over until the T on the stator and the line on the rotor are in alignment. The pistons are now on TDC, and one cylinder will be on the compression stroke. This can be verified by feeling for free play at the rocker arms. Check the clearances between the rocker arms and valves on that cylinder, using feeler gauges. If the clearance is not correct, slacken the locknut and turn the square-headed adjuster screw to obtain the specified setting.

Rotate the engine through 360°, and repeat the procedure on the other cylinder. The clearances should be as follows; measured on a cold engine:

Valve clearances (cold engine)

XS1B, XS 2 TX 650	Inlet Exhaust	0.15 mm 0.30 mm	(0.006 in) (0.012 in)
TX 650A XS 650B	Inlet Exhaust	0.05 mm 0.10 mm	(0.002 in) (0.004 in)
XS 650C	Inlet Exhaust	0.05 mm 0.15 mm	(0.002 in) (0.006 in)

12 Checking and adjusting the cam chain tension

1 Remove both spark plugs from the cylinder head and rotate the engine forwards (anti-clockwise viewed from the left-hand side of the machine) so that the slack in the cam chain is placed in the rear run of the chain. Remove the cap nut from the chain tensioner assembly located to the rear of the cylinder barrel and check that the end of the tensioner plunger is flush with the outer adjuster bolt. If not, turn the adjuster bolt until the end of the plunger is flush. Rotate the engine forwards to check that the setting remains correct and lock the adjuster bolt by refitting the cap nut. Refit the spark plugs.

nipple

inner cable

plasticine funnel around outer cable

cable suspended vertically

cable lubricated when oil drips from far end

Fig. 1 Oiling a control cable

RM 7 Remove and clean the oil filter element

RM 8 ... and also the sump filter element

Grease the swinging arm pivot shaft

Check and set valve clearances

Six monthly or every 4000 miles

Complete the checks listed under the weekly/250 mile and three monthly/2000 mile headings, then complete the following additional procedures:

1   Changing the front fork damping oil

Place the machine on the centre stand so that the front wheel is clear of the ground. Place wooden blocks below the crankcase in order to prevent the machine from tipping forward. Loosen and remove the chrome cap bolts. Unscrew the drain plug from each fork leg, located directly above the wheel spindle, and allow the damping fluid to drain into a suitable container. This is accomplished most easily if the legs are attended to in turn. Take care not to spill any fluid onto the brake disc or tyre. The forks may be pumped up and down slowly to expel any remaining fluid. Refit and tighten the drain plugs. Refill each fork leg with SAE 10W/30 engine oil, or a good quality fork oil. If a straight grade fork oil is chosen, either SAE 10, 20 or 30 may be used, depending on choice. The thicker the oil the heavier will be the damping. Refit and tighten the chrome cap bolts.

The front fork capacities are as follows:

XS1B	223 cc	per leg
XS2, TX650	136 cc	per leg
TX650A, XS650B XS650C	155 cc	per leg

General maintenance adjustments

1   Clutch adjustment

The intervals at which the clutch should be adjusted will depend on the style of riding and the conditions under which the machine is used.

Adjust the clutch in two stages as follows:

Remove the clutch adjuster cover, which is retained by two screws. Loosen the cable adjuster screw locknut and turn the adjuster inwards fully, to give plenty of slack in the inner cable. Loosen the adjuster screw locknut in the casing and turn the screw clockwise until slight resistence is felt. Back off the screw about ¼ turn and tighten the locknut. The cover may now

be replaced.

Undo the cable adjuster screw at the handlebar lever, until there is approximately 2 - 3 mm (0.08 - 0.12 in) play measured between the inner face of the lever and the stock face. Finally, tighten the cable adjuster locknut.

2   Disc pads - checking

The rate at which disc pads wear is also largely dependent on both the owner's riding style and also the type of journey normally undertaken. It follows that a machine used for frequent town work will wear pads out at a greater rate than a machine normally used for touring.

Pad wear can be checked very easily by measuring the small gap between the disc surface and the tang which projects from the pad. If this is less than 0.5 mm (0.002 in), the pads should be removed and renewed. Chapter 5 Section 3 should be consulted for details.

3   Brake linings - checking and adjustment

The front drum brake (early models), and rear drum brake (all models) linings should be checked for wear periodically. With the relevant wheel removed, and the brake plate withdrawn, measure the overall diameter across the shoes. The dimensions are as follows:

Overal diameter of brake shoes

	Nominal	Wear limit
Front	192 mm (7.556 in)	188 mm (7.400 in)
Rear	172 mm (6.772 in)	168 mm (6.612 in)

Lining thickness, front and rear

Nominal	Wear limit
4 mm (0.160 in)	2 mm (0.080 in)

Renew the shoes if they are contaminated with oil or grease (having first eliminated the source of the leak) or if the measurements are at or below the wear limit.

Note that later models incorporate a switch in the rear brake drum which is operated by a cam on the fulcrum spindle. The switch operates a warning light which indicates that the shoes are nearing the wear limit.

On reassembly, adjust the brakes so that the wheel is just free to spin when the brake lever or pedal is released. On the front drum brake, allow about 2-3 mm of slack in the operating cable, and a similar margin in the rear brake rod. The actual point at which a brake commences operation is largely discretionary.

Set clutch free play with this adjuster

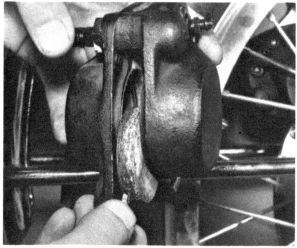

Fit new pads if old ones are badly worn

Dimensions and weight

Dimensions and weight

XS 1, XS 1B

Overall length	85.4 in (2170 mm)
Overall width	35.6 in (905 mm)
Overall height	45.3 in (1150 mm)
Wheelbase	55.5 in (1410 mm)
Ground clearance...	5.9 in (150 mm)
Net weight: XS 1	439 lbs
XS 1B	409 lbs

XS 2, TX 650

Overall length	85.6 in (2175 mm)
Overall width	35.6 in (905 mm)
Overall height	45.9 in (1165 mm)
Wheelbase...	55.5 in (1410 mm)
Ground clearance	5.9 in (150 mm)
Net weight...	427 lbs (194 kg)

TX 650A/XS 650B, XS 650C

Overall length	85.8 in (2180 mm)
Overall width	35.4 in (907 mm) (32.9 in/835 mm XS 650C)
Overall height	45.7 in (1162 mm) (44.9 in/1140 mm XS 650C)
Wheelbase	56.5 in (1435 mm)
Ground clearance...	5.5 in (140 mm)
Net weight	467 lbs (212 kgs)

Quick Glance
maintenance adjustments and capacities

Engine/gearbox unit	Top up with SAE 20W/50 engine oil as required
Contact breaker gap	0.3 – 0.45 mm (0.012 – 0.018 in)

Valve clearances			
	XS1B, XS2 & TX 650	Inlet Exhaust	0.15 mm (0.006 in) 0.30 mm (0.012 in)
	TX650A, XS650B	Inlet Exhaust	0.05 mm (0.002 in) 0.10 mm (0.004 in)
	XS650C	Inlet Exhaust	0.05 mm (0.002 in) 0.15 mm (0.006 in)

Sparking plug gap	0.6 – 0.7 mm (0.024 – 0.028 in)
Tyre pressures	Front 23 - 25 psi (1.6 kg/cm^2) Rear 28 - 30 psi (2.0 kg/cm^2)

Recommended lubricants

Component	Viscosity/product	Model	Quantity
Engine/gearbox unit	SAE 20W/50 motor oil	XS1, XS1B	: 3.2 litres (5.6 imp pints 3.4 US qts)
		XS2, TX650	: 2.6 litres (4.6 imp pints 2.7 US qts)
		TX650A, XS650B XS650C	: 2.5 litres (4.4 imp pints 2.6 US qts)
Front forks	Fork oil or SAE 10W/30 motor oil	XS1, XS1B	: 223 cc per leg
		XS2, TX650	: 136 cc per leg
		TX650A, XS650B XS650C	: 155 cc per leg
Disc brake	Hydraulic fluid to SAE J1703B	as required	
Greasing points	High melting point grease	as required	
Final drive chain	Chain lubricant	as required	

Working conditions and tools

When a major overhaul is contemplated, it is important that a clean, well-lit working space is available, equipped with a workbench and vice, and with space for laying out or storing the dismantled assemblies in an orderly manner where they are unlikely to be disturbed. The use of a good workshop will give the satisfaction of work done in comfort and without haste, where there is little chance of the machine being dismantled and reassembled in anything other than clean surroundings. Unfortunately, these ideal working conditions are not always practicable and under these latter circumstances when improvisation is called for, extra care and time will be needed.

The other essential requirement is a comprehensive set of good quality tools. Quality is of prime importance since cheap tools will prove expensive in the long run if they slip or break when in use, causing personal injury or expensive damage to the component being worked on. A good quality tool will last a long time, and more than justify the cost.

For practically all tools, a tool factor is the best source since he will have a very comprehensive range compared with the average garage or accessory shop. Having said that, accessory shops often offer excellent quality tools at discount prices, so it pays to shop around. There are plenty of tools around at reasonable prices, but always aim to purchase items which meet the relevant national safety standards. If in doubt, seek the advice of the shop proprietor or manager before making a purchase.

The basis of any tool kit is a set of open-ended spanners, which can be used on almost any part of the machine to which there is reasonable access. A set of ring spanners makes a useful addition, since they can be used on nuts that are very tight or where access is restricted. Where the cost has to be kept within reasonable bounds, a compromise can be effected with a set of combination spanners – open-ended at one end and having a ring of the same size on the other end. Socket spanners may also be considered a good investment, a basic $3/8$ in or $1/2$ in drive kit comprising a ratchet handle and a small number of socket heads, if money is limited. Additional sockets can be purchased, as and when they are required. Provided they are slim in profile, sockets will reach nuts or bolts that are deeply recessed. When purchasing spanners of any kind, make sure the correct size standard is purchased. Almost all machines manufactured outside the UK and the USA have metric nuts and bolts, whilst those produced in Britain have BSF or BSW sizes. The standard used in USA is AF, which is also found on some of the later British machines. Others tools that should be included in the kit are a range of crosshead screwdrivers, a pair of pliers and a hammer.

When considering the purchase of tools, it should be remembered that by carrying out the work oneself, a large proportion of the normal repair cost, made up by labour charges, will be saved. The economy made on even a minor overhaul will go a long way towards the improvement of a toolkit.

In addition to the basic tool kit, certain additional tools can prove invaluable when they are close to hand, to help speed up a multitude of repetitive jobs. For example, an impact screwdriver will ease the removal of screws that have been tightened by a similar tool, during assembly, without a risk of damaging the screw heads. And, of course, it can be used again to retighten the screws, to ensure an oil or airtight seal results. Circlip pliers have their uses too, since gear pinions, shafts and similar components are frequently retained by circlips that are not too easily displaced by a screwdriver. There are two types of circlip pliers, one for internal and one for external circlips. They may also have straight or right-angled jaws.

One of the most useful of all tools is the torque wrench, a form of spanner that can be adjusted to slip when a measured amount of force is applied to any bolt or nut. Torque wrench settings are given in almost every modern workshop or service manual, where the extent to which a complex component, such as a cylinder head, can be tightened without fear of distortion or leakage. The tightening of bearing caps is yet another example. Overtightening will stretch or even break bolts, necessitating extra work to extract the broken portions.

As may be expected, the more sophisticated the machine, the greater is the number of tools likely to be required if it is to be kept in first class condition by the home mechanic. Unfortunately there are certain jobs which cannot be accomplished successfully without the correct equipment and although there is invariably a specialist who will undertake the work for a fee, the home mechanic will have to dig more deeply in his pocket for the purchase of similar equipment if he does not wish to employ the services of others. Here a word of caution is necessary, since some of these jobs are best left to the expert. Although an electrical multimeter of the AVO type will prove helpful in tracing electrical faults, in inexperienced hands it may irrevocably damage some of the electrical components if a test current is passed through them in the wrong direction. This can apply to the synchronisation of twin or multiple carburettors too, where a certain amount of expertise is needed when setting them up with vacuum gauges. These are, however, exceptions. Some instruments, such as a strobe lamp, are virtually essential when checking the timing of a machine powered by CDI ignition system. In short, do not purchase any of these special items unless you have the experience to use them correctly.

Although this manual shows how components can be removed and replaced without the use of special service tools (unless absolutely essential), it is worthwhile giving consideration to the purchase of the more commonly used tools if the machine is regarded as a long term purchase Whilst the alternative methods suggested will remove and replace parts without risk of damage, the use of the special tools recommended and sold by the manufacturer will invariably save time.

Chapter 1 Engine, clutch, and gearbox

Refer to Chapter 7 for details of 1977 to 1983 models

Contents

Specification

Engine

Type ...	Parallel twin cylinder, air cooled four stroke
Capacity ...	653 cc (39.85 cu in) all models
Bore ...	75 mm (2.953 in) all models
Stroke ...	74 mm (2.913 in) all models
Power output...	53 bhp @ 7000 rpm - XSIB, XS2, TX 650
	43.36 bhp @ 7000 rpm - TX 650A
	45.61 bhp @ 7500 rpm - XS 650C
Torque...	40.1 ft lbs @ 6000 rpm - XSIB, XS2, TX650
	36.01 ft lbs @ 5500 rpm - TX 650A
	36.16 ft lbs @ 6000 rpm - XS 650C

Crankshaft assembly

Small end bearings

Type ...	... ...	Needle roller
Maximum side play ...	... ...	2 mm (0.079 in)
Maximum axial play ...	... ...	2 mm (0.079 in)
Nominal axial play ...	... ...	1 mm (0.039 in)

Big end bearings:

Type	... ...	Roller
Connecting rod to flywheel	... ...	
clearance at big end eye ...	... ...	0.3 - 0.6 mm (0.012 - 0.024 in)
Maximum runout at journals ...	... ...	0.03 mm (0.0012 in)

Cylinders

Type	... ...	Light alloy cylinder block with steel liners
Nominal bore diameter ...	... ...	75.0 mm (2.960 in)
Wear limit	... ...	75.1 mm (2.964 in)
Nominal bore taper ...	... ...	0.005 mm (0.0002 in)
Wear limit	... ...	0.050 mm (0.002 in)

Pistons

Type	... ...	Light alloy, floating gudgeon pin
Available sizes ...	... ...	Std, +0.25 mm, +0.50 mm, +0.75 mm, +1.0 mm
Nominal piston to bore clearance ...	... ...	0.050 - 0.055 mm (0.002 - 0.0022 in)
Wear limit	... ...	0.100 mm (0.0039 in)

Piston rings

Type

Top ...	... ...	Plain, compression
Middle ...	... ...	Tapered, wiper
Bottom ...	... ...	Double rail oil scraper, with expander

End gap in bore

Top ...	... ...	0.2 - 0.4 mm (0.008 - 0.016 in), wear limit 0.8 mm (0.031 in)
Middle ...	... ...	0.2 - 0.4 mm (0.008 - 0.016 in), wear limit 0.8 mm (0.031 in)
Bottom ...	... ...	0.3 - 0.6 mm (0.012 - 0.024 in), wear limit 1.0 mm (0.040 in)

Ring to land clearance

Top ...	... ...	0.04 - 0.08 mm (0.0016 - 0.0031 in), wear limit 0.15 mm (0.006 in)
Middle ...	... ...	0.04 - 0.08 mm (0.0016 - 0.0031 in), wear limit 0.15 mm (0.006 in)
Bottom ...	... ...	Nil

Cylinder head

Type	... ...	Light alloy, carrying camshaft and two valves per cylinder

Camshaft

Type	... ...	Single overhead camshaft
Bearings ...	... ...	Four single-row ball races
Sprocket ...	... ...	34 tooth
Drive chain ...	... ...	Single row, endless type

Cam lift:	XS1B, XS2, TX650	TX650A, XS650B, XS650C
Inlet	7.44 mm (0.290 in)	7.991 mm (0.314 in)
Exhaust	7.12 mm (0.278 in)	8.030 mm (0.316 in)
Cam height:		
Inlet	39.63 ± 0.05 mm (1.545 ± 0.0019 in)	39.99 ± 0.05 mm (1.574 ± 0.0019 in)
Exhaust	39.36 ± 0.05 mm (1.535 ± 0.0019 in)	40.03 ± 0.05 mm (1.576 ± 0.0019 in)

Rocker arms and shafts

Rocker bore diameter ...	... ...	15.01 mm (0.59 in) nominal
Rocker shaft diameter ...	... ...	14.98 mm (0.589 in) nominal

Valves

Inlet valve stem diameter ...	... ...	7.975 - 7.990 mm (0.3140 - 0.3148 in)
Exhaust valve stem diameter ...	... ...	7.960 - 7.975 mm (0.3134 - 0.3140 in)
Inlet valve guide bore diameter ...	... ...	8.010 - 8.019 mm (0.3150 - 0.3157 in)
Exhaust valve guide bore diameter ...	... ...	8.010 - 8.019 mm (0.3150 - 0.3157 in)

Valve guides available in two outside diameter oversizes

	Nominal	Wear limit
Valve stem to guide bore clearances - Inlet	0.020 - 0.044 mm (0.0008 - 0.0017 in)	0.1 mm (0.0039 in)
Exhaust	0.035 - 0.059 mm (0.0014 - 0.0023 in)	0.12 mm (0.0047 in)

Valve seat angle	... ...	45°
Valve seat width	... ...	1.3 mm (0.051 in) wear limit 2.0 mm (0.078 in)

Valve timing

XSIB, XS2, TX 650 models

Inlet opens	47° BTDC	⎫	Valve open for 294°
Inlet closes	67° ABDC	⎬	
Exhaust opens	60° BBDC	⎫	Valve open for 281°
Exhaust closes	41° ATDC	⎭	

TX 650A, XS 650B, XS 650C models

Inlet opens	36° BTDC	}	Valve open for 284°
Inlet closes	68° ABDC	}	
Exhaust opens	68° BBDC	}	Valve open for 284°
Exhaust closes	36° ATDC	}	

Valve springs

	nominal	wear limit
Free length		
Inner	41.0 mm (1.614 in)	39.0 mm (1.535 in)
Outer	41.8 mm (1.646 in)	(39.8 mm 1.567 in)

Clutch

Type	Wet, multi-plate
No of plates:	
XS1, XS1B and XS2	6 friction, 5 plain
TX650 and all XS650 models	7 friction, 6 plain
Spring free length	34.6 mm (1.363 in)
Wear limit	33.6 mm (1.324 in)
Spring rate	2.6 kg/mm (145.592 lb/in)
Friction plate thickness	3.0 mm (0.118 in)
Wear limit	2.7 mm (0.106 in)
Warp limit	0.05 mm (0.002 in)

Gearbox

Type	Constant mesh, 5-speed, wide ratio
Reduction ratios:	
1st	2.461 : 1 (32/13)
XS1 and XS1B	2.214 : 1 (31/14)
2nd	1.588 : 1 (27/17)
3rd	1.300 : 1 (26/20)
4th	1.095 : 1 (23/21)
5th	0.956 : 1 (22/23)
Primary drive ratio	2.666 : 1 (72/27)
Final drive ratio	2.000 : 1 (34/17)

Torque settings

Cylinder head; 10 mm nuts :	22 - 25 ft - lbs (3.0 - 3.5 m - kgs)
6 mm bolt :	7.5 - 11 ft - lbs (1.0 - 1.5 m - kgs)
8 mm bolts :	15 - 18 ft - lbs (2.1 - 2.5 m kgs)
10 mm bolts :	11 - 14.5 ft - lbs (1.5 - 2.0 m - kgs)
Engine mounting nuts	25 - 35 ft - lbs (3.5 - 4.8 m - kgs)
Oil filter cover screws	6.0 - 7.2 ft - lbs (0.8 - 1.0 m - kgs)
Oil feed union : 10 mm bolts :	14.5 - 16 ft - lbs (2.0 - 2.2 m - kgs)
14 mm bolts :	18.0 - 22.0 ft - lbs (2.5 - 3.0 m - kgs)
Crankcase drain plug	25 - 29 ft - lbs (3.5 - 4.0 m - kgs)
Kickstarter pinch bolt	11 - 18 ft - lbs (1.5 - 2.5 m - kgs)
Alternator rotor nut	50 - 54 ft - lbs (7.0 - 7.5 m - kgs)
Alternator stator screws:	5.0 - 6.5 ft - lbs (0.7 - 0.9 m - kgs)
Clutch centre nut...	54 - 58 ft - lbs (7.5 - 8.0 m - kgs)
Gearbox sprocket nut	72 - 87 ft - lbs (10.0 - 12.0 m - kgs)
Crankcase bolts and nuts	14.5 ft - lbs (2.0 m - kgs)

1 General description

The engine fitted to the Yamaha 650 twins is a single overhead camshaft vertical twin. A 360° crankshaft is carried in three main bearings.

A centrally placed sprocket drives the single overhead camshaft, by way of a single-row chain which runs in a tunnel between the cylinders. The various engine castings are of light alloy, the block being fitted with dry steel liners. The crankcases, which house the crankshaft assembly and gearbox components, separate horizontally to make dismantling easier, and to preserve oil-tightness in the wet sump engine unit. Engine lubrication is provided by a trochoidal oil pump, giving a pressure feed to the main components. The gearbox internals run in the oil bath formed by the lower crankcase half. A multi-plate clutch, mounted on the left-hand side of the unit, transmits power from the engine to the gearbox via a gear primary drive. Secondary drive is by way of a heavy roller chain to the rear wheel.

Starting is provided by an electric starter motor housed on the underside of the unit, and supplemented by a kickstart pedal for emergency use. Power for the electrical system is derived from a crankshaft-mounted alternator, in conjunction with a twelve volt battery. The ignition system is of the conventional coil and contact breaker type.

2 Operations with the engine/gearbox unit in the frame

It is not necessary to remove the unit from the frame in order to carry out the following operations:-
1 Removing the right and left-hand outer covers.
2 Dismantling the clutch assembly and primary drive
3 Removing and replacing the oil filter element
4 Removing and replacing the alternator
5 Removing and replacing the starter motor
6 Removing and replacing the gearbox sprocket
7 Removing the gear selector claw and shaft and stopper mechanism
8 Removing and replacing the carburettors

3 Operations with engine/gearbox unit removed from frame

As previously described the crankshaft and gearbox assemblies are housed within a common casing. Any work carried out on either of these two major assemblies will necessitate removal of the engine from the frame so that the crankcases can be separated.

4 Removing the engine/gearbox unit

1   Run the engine until normal operating temperature is achieved. This will thin the oil and assist in draining the sump. Place the machine firmly on its centre stand, ensuring that it is in no danger of toppling over during the removal sequences. It is helpful, but by no means essential, to stand the machine on a raised platform, constructed with planks supported firmly by blocks. With the machine raised by about two feet, it will be found that working is much more convenient.

2   Place a tray or bowl of at least 3.2 litres (5.6 Imp pints, 3.4 US quarts) beneath the engine unit and remove the two large drain plugs. The unit should be left to drain thoroughly before refitting the plugs. Lift the dualseat and disconnect the battery positive (+) lead. Alternatively, the battery can be removed completely and placed to one side.

3   Ensure that both fuel taps are in the off position, then remove the petrol feed pipes. Unscrew the retaining bolt at the rear of the fuel tank, and pull it up and back to disengage the front mounting rubbers. Place the tank in a safe place, to await reassembly.

4   Remove both side panels by turning the catches and lifting them up and clear of the locating tangs on the frame. On early models, the panels are retained by pinned bolts. Disconnect the alternator cables at the multi-way connector which is located adjacent to the air filter.

5   Unhook the throttle cable(s) and place them out of the way by laying them along the top tube. Slacken the clips which retain the air hoses to the carburettors. Remove both air cleaner assemblies by unscrewing the two top bolts immediately behind the seat mounting stud, and the bolt at each side.

6   Slacken the worm drive clips which hold the carburettors to the rubber inlet stubs, and pull the carburettors and support bar free. Using a socket screw key, remove one of the rubber inlet stubs. This will enable the carburettors to be withdrawn as an assembly.

7   Disconnect the neutral indicator switch lead from the spring-loaded terminal pin. The switch is screwed into the left-hand crankcase at the rear of the unit. Detach the sparking plug leads and place them out of harm's way by lodging the suppressor caps on the top frame tube. Disconnect the leads from the contact breakers at the connector block adjacent to the coils. On early electric start models, disconnect the decompressor cable. Detach the leads to the horn and remove it, after removing the two mounting bolts.

8   Remove the left-hand outer casing, having first detached the gearchange pedal, and slackened the casing securing screws. As the cover is lifted away, disengage the clutch cable from the

operating mechanism. It may prove necessary to detach the cable from the handlebar lever to obtain sufficient slack.

9   Knock back the gearbox sprocket tab washer. Slacken the gearbox sprocket nut, having locked the sprocket by applying the rear brake (Note that whilst it may not prove necessary to remove the sprocket, it is much easier to do so at this stage rather than when the engine unit has been removed from the frame).

10 Turn the rear wheel until the joining link is positioned on the rear sprocket. Prise off the spring link and slide the chain apart, reassembling the link on one end of the chain to avoid subsequent loss. The chain can now be removed.

11 Slacken the pinch bolt which clamps the exhaust system balance pipe between the two silencers. The complete exhaust system should now be removed from the machine . The exhaust pipes are each retained at the exhaust ports by a flange fitting, and by a single bolt just forward of the silencers. The silencers are held by two mounting bolts each. Place the exhaust system in a place of safety, to avoid accidental damage to the chromium plating.

12 Remove the two nuts holding each of the rubber-mounted footrest bars to the frame. Slide each assembly off the mounting studs and place to one side. Slacken the two bolts which retain the head steady plates to the cylinder head, followed by the nuts holding the plates to the frame. Note that it is not necessary to dismantle the head steady completely. The two plate halves can be swung upwards and lodged against the frame top tube, with the condensers left in position.

13 Slacken the pinch bolt retaining the rear brake pedal, and remove it, to allow more room to manoeuvre the unit out of the frame. The kickstart can be left in position as it is useful as a lifting point. Remove the tachometer drive cable.

14 Remove the engine front mounting bolts, and detach the engine plates from the frame. The two engine upper rear mounting bolts should be removed next, followed by the engine lower rear mounting bolts. The engine is now retained by the central mounting bolt only.

15 Whilst it is not absolutely essential, it is recommended that the starter motor is removed at this stage. It will be noted that the starter motor terminal is located immediately adjacent to the frame, and is easily damaged during removal. The motor is held in position by four bolts which pass through lugs into the crankcase. The engine unit can be rocked on its remaining mounting bolt to make access easier.

16 Two people will be needed to lift the unit from the frame as the last bolt is removed. It is not advisable to attempt this stage unaided as there is a risk of damage to the paintwork, and it is also very easy to topple the machine. The unit may be removed from the right or left-hand side, and placed on a suitable work-bench to await further dismantling.

4.2 Disconnect battery terminals and remove battery   4.3 Tank is retained by single bolt at rear

4.5a Disconnect throttle cable(s)

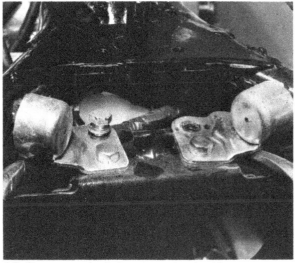

4.5b Remove securing bolts at top and at the side ...

4.5c ... and withdraw air cleaner assemblies

4.6a Detach one of the inlet stubs, to allow ...

4.6b ... carburettors to be removed as an assembly

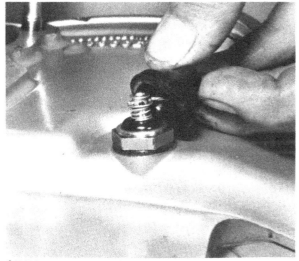

4.7 Release neutral switch lead

4.8 Release clutch cable from operating mechanism

4.9 Apply rear brake, and slacken gearbox sprocket nut

4.11a Slacken balance pipe clamp

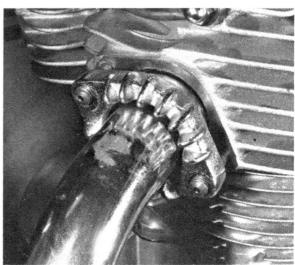

4.11b Pipes are each retained by a clamp

4.12 Remove cylinder head steady plates

4.13 Disconnect tachometer drive cable at gland nut

4.14a Remove engine front bolts and plates ...

4.14b ...followed by rear upper mounting bolts and plates,...

4.14c ... and single lower bolt

4.15 It is advisable to remove the starter motor

4.16a Remove the remaining bolt

4.16b Engine can now be lifted out of frame

5 Dismantling the engine/gearbox unit: general

1 Before commencing work on the engine unit, the external surfaces should be cleaned thoroughly. A motorcycle has very little protection from road grit and other foreign matter which sooner or later will find its way into the dismantled engine if this simple precaution is not carried out.
2 One of the proprietary cleaning compounds such as Gunk or Jizer can be used to good effect, especially if the compound is first allowed to penetrate the film of grease and oil before it is washed away. In the USA, Gumout degreaser is an alternative.
3 It is essential when washing down to make sure that water does not enter the carburettors or the electrics particularly now that these parts are more vulnerable.
4 Collect together an adequate set of tools in addition to those of the tool roll carried under the seat.
5 If the engine has not been previously dismantled, an impact screwdriver will prove essential. This will safeguard the heads of the crosshead screws used for engine assembly. These are invariably machine-tightening during manufacture. **Caution** - Use great care as the screws and cases are easily damaged. Use a crosshead type screwdriver and NOT one of the Phillips type, which will slip out of the screws.
6 Avoid force in any of the operations. There is generally a good reason why an item is difficult to remove, probably due to the use of the wrong procedure or sequence of operations.
7 Dismantling will be made easier if a simple engine stand is constructed that will correspond with the engine mounting points. This arrangement will permit the complete unit to be clamped rigidly to the workbench, leaving both hands free for dismantling.

6 Dismantling the engine/gearbox: removing the cylinder head cover and and cylinder head

1 Slacken the gland nut and banjo bolts which retain the oil feed pipe to the crankcase and cylinder head cover respectively, and lift the pipe away. Remove the contact breaker cover. It is retained by two screws on the left-hand end of the camshaft.
2 Use a centre punch to mark the position (A) of the contact breaker base plate in relation to the contact breaker housing, If this precaution is not taken, it will be necessary to reset the ignition timing on reassembly. Remove the two large slotted screws (B) to release the complete contact breaker and base plate assembly. The contact breaker housing can now be removed after releasing the three crosshead screws which retain it to the cylinder head cover.
3 Remove the cover from the opposite end of the camshaft to expose the auto-advance unit. Slacken the central retaining nut and remove it, along with the slotted boss which conveys movement from the bob-weights to the contact breaker assembly.

The advance rod can be withdrawn from the contact breaker end of the camshaft. Alternatively, the nut can be removed from the contact breaker end of the rod, as the contact breaker assembly is removed. The rod can then be withdrawn from the auto-advance end, complete with the slotted boss, It is advisable to reassemble the various components in their normal positions on the rod, to avoid any subsequent confusion. Prise off the clips which retain the bob-weights, and slide them off their pivot posts.
4 The auto-advance unit can be removed after the slotted nut which retains it has been tapped free, using a small punch and hammer. A small pin will be found at the end of the camshaft, which drives the advance unit. It is necessary to remove this pin in order that the auto-advance housing can be removed. If it is found impossible to dislodge it, using pointed-nosed pliers, remove the three housing retaining screws, and leave the housing loose on the camshaft. It can then be removed after the cylinder head cover has been lifted off. Remove the valve adjustment covers.
5 The cylinder head cover itself is retained by a total of twelve bolts and nuts which also retain the cylinder head. In addition to these, there are three further bolts which pass through the cylinder head only. These are located adjacent to the sparking plug holes and at the front of the head between the cylinders. In order to obviate any risk of cylinder head warpage, remove all fifteen of the domed nuts and bolts., following the slackening sequence shown in Fig 1.15. The cylinder head cover can now be lifted off. If the auto-advance housing is still in position due to the driving pin having proved difficult to remove, it can be disengaged from the camshaft, taking care not to damage the seal. Note that bolts 12, 13, 14 and 15 in Fig. 1.1 are fitted with rubber-sealing washers.
6 Slacken and remove the six bolts holding the cam chain tensioner in position. Note that the top right-hand and lower left-hand ones are dowel bolts. The tensioner assembly can be pulled clear of the cylinders and placed to one side, to await reassembly.
7 Turn the engine slowly until the riveted link of the cam chain is at the top of the camshaft sprocket. It can be identified by the punch marks, in the form of slots, across the rivet heads. Borrow the appropriate tool with which to separate the chain. This takes the form of an anvil and threaded punch which can be screwed inwards to push out the rivets. If this is not available, a nut splitter can be modified, with a little ingenuity, and used to good effect. This method is shown in the accompanying photograph. Unless the engine is to be completely dismantled, take care that the chain does not drop down into the crankcase. The ends should be wired to a convenient stud, to avoid this possibility. The camshaft can now be lifted clear of the cylinder head.
8 The cylinder head can be lifted off the cylinder block, taking care not to let the cam chain drop into the crankcase. The wire holding the ends of the chain should be secured after the cylinder head is removed.

Fig. 1.1 Cylinder head and cylinder head cover (see opposite page)

1 Cylinder head assembly	11 Bolt - 4 off	21 Stud - 4 off	31 Cover - 2 off
2 O ring - 4 off	12 Plain washer - 4 off	22 Plain washer - 4 off	32 Screw - 4 off
3 Inlet valve guide - 2 off	13 Inspection cover - 3 off	23 Domed nut - 4 off	33 Rubber blank
4 Exhaust valve guide - 2 off	14 O ring - 3 off	24 O ring - 2 off	34 Spark plug - 2 off
5 Dowel pin - 2 off	15 Stud - 3 off	25 Gasket - 2 off	35 Breather assembly
6 O ring - 4 off	16 Stud - 6 off	26 Contact breaker/ATU unit	36 Gasket
7 Sleeve - 4 off	17 Plain washer - 9 off	housing - 1 each	37 Socket screws - 4 off
8 Domed nut - 8 off	18 Domed nut - 9 off	27 Plain washer - 2 off	38 Breather pipe - 2 off
9 Washer - 4 off	19 Inspection cover	28 Screw - 2 off	39 Protection spring - 2 off
10 Sealing washer - 4 off	20 O ring	29 Shakeproof washer - 29	40 Spring clip - 4 off
		30 Countersunk screw - 6 off	41 Plate

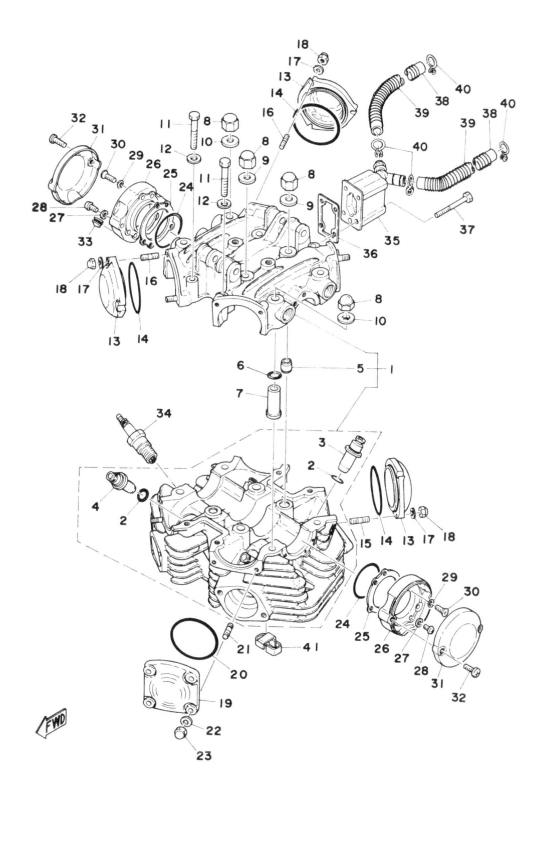

Fig.1.1 Cylinder head and cylinder head cover

6.1a Remove oil feed gland nut, and ...

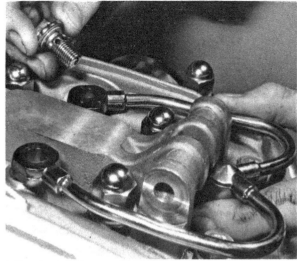

6.1b ... banjo bolts, to release oil feed assembly

6.2a Centre punch at A to retain timing. Remove screws B and ...

6.2b ...withdraw contact breaker base plate

6.3a Shaft can be moved from either end of camshaft

6.3b Remove ATU bobweights from pivot pins

6.4a Baseplate is retained by slotted nut

6.4b Remove baseplate, followed by ...

6.4c ... driving pin (leave pin in position if very tight)

6.4d Detach valve rocker inspection covers

6.5a Do not omit these bolts adjacent to sparking plugs ...

6.5b ... and single rear bolt

6.5c Disengage ATU housing if still on camshaft end

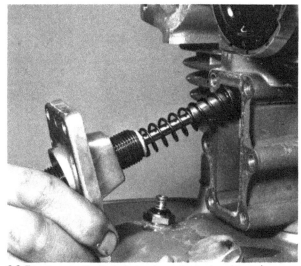

6.6 Remove camshaft chain tensioner (note dowel bolts)

6.7a Note joining link marked with punch lines

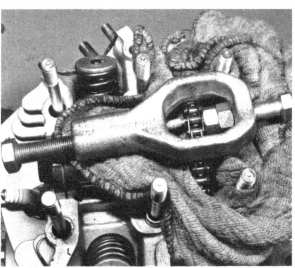

6.7b Use chain breaker or improvised tool to drive out link

6.7c Secure ends of chain with wire, and remove camshaft

6.8 Remove cylinder head, taking care not to drop chain

7 Dismantling the engine/gearbox: removing the cylinder block, pistons and rings

1   Rotate the crankshaft until both pistons are at TDC. With the use of a soft-nosed mallet free the cylinder block and work it gently up the holding down studs. If the engine has not been dismantled since leaving the factory, the cylinder block may effectively be glued to the crankcase mouth due to the type of sealing compound applied to the base gasket.
2   Slide the cylinder block up and off the pistons, taking care to support each piston as the cylinder block becomes free. If a top and overhaul only is being carried out, place a clean rag in each crankcase mouth before the lower edge of each cylinder frees the rings. This will preclude any small particles of broken ring falling into the crankcase.
3   Prise the outer gudgeon pin circlip of each piston from position. The gudgeon pins are a light push fit in the piston bosses so can be removed with ease. If any difficulty is encountered, apply to the offending piston crown a rag over which boiling water has just been poured. This will give the necessary temporary expansion to the piston bosses to allow the gudgeon pin to be pushed out. Before removing each piston, scribe the cylinder identification inside the piston skirt. A mark R or L will ensure that the piston is replaced in the correct bore, on reassembly. It is unnecessary to mark the back and front of the piston because this is denoted by an arrow mark cast in the piston crown.
4   Each piston is fitted with two compression rings and an oil control ring. It is wise to leave the rings in place on the pistons until the time comes for their examination or renewal, in order to avoid confusing their correct order.
5   The cam chain guide may be removed from the cylinder block, after releasing the two bolts which retain it. It is not necessary to disturb the outer, chromium plated, nuts through which these pass. The chain tensioner pivots on a baseplate which is screwed between the crankcase mouths. If it is desired that the assembly be removed, take care that none of the retaining screws are dropped into the crankcase.

8 Dismantling the engine/gearbox: removing the left-hand crankcase cover and alternator

1   Remove the six outer casing screws, and lift the cover away. Trace the alternator wiring from the stator, and disengage it from the guide attached to the crankcase. Slacken the two long crosshead screws which hold the alternator stator to the crankcase, and lift the stator away. Note the small locating pin at the bottom of the stator. If this is loose, it should be removed for safe keeping.
2   The rotor is a built-up unit keyed to a taper on the mainshaft. It is recommended that the official Yamaha extractor is used to remove the rotor, to obviate any risk of damage. It is possible to utilise a conventional 3-legged puller, providing great care is taken when using it. The legs must be arranged to bear on the inner segments of the rotor. If this is not done, there will be a real danger of pulling the assembly apart. Slacken the centre nut, having locked the crankshaft by passing a bar through one of the connecting rod eyes, and resting the ends on wooden blocks at each side of the crankcase mouth. The extractor can now be fitted, and the rotor drawn off its taper.

9 Dismantling the engine/gearbox: removing the gearbox final drive sprocket

1   If the sprocket securing nut has not been slackened during the engine removal sequences, it will be necessary to lock the engine to do so. This can be accomplished by arranging the crankshaft as described in the preceding Section, and selecting top gear by temporarily refitting the gearchange pedal.
2   Knock back the tab washer and slacken the securing nut. The sprocket can now be removed from the mainshaft.

10 Dismantling the engine/gearbox: removing the oil filter

1   The oil filter element is housed in a compartment at the front of the right-hand outer cover, and can be removed after unscrewing the two screws which retain the circular cover plate. The housing should be carefully cleaned out prior to reassembly and the gauze element washed in petrol and dried, to remove any accumulated grime.

11 Dismantling the engine/gearbox: removing the primary drive cover and dismantling the clutch

1   The primary drive cover is retained by ten (10) crosshead screws. Remove the screws followed by the casing and the gasket.
2   Unscrew the clutch bolts and remove them together with the springs and washers. Remove the clutch pressure plate.
3   Lift out the clutch plates making sure they remain in their correct relative positions. Pull out the mushroom head clutch pushrod and the main length of the clutch pushrod from the left-hand side. Note that a steel ball lies between the two pushrods.
4   The clutch centre is retained by a nut, plus one plain and one dished washer. It will be necessary to lock the clutch centre to enable the nut to be removed. This is best done using a clutch holding tool, which consists of a plain clutch plate with a length of rod welded on as a handle, and is available as a Yamaha Service tool. Alternatively, if the tool is not available, it may be possible to use steel strips to sprag the centre, (taking care that the centre or drum are not damaged) or to hold the centre with a strap wrench.
5   With the clutch centre held securely, slacken and remove the centre nut and washers. The clutch centre can now be lifted off, followed by the thrust plate and bearing. The clutch drum should be removed next, together with its bearing sleeve and washers.

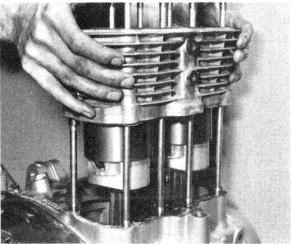

7.2 Remove cylinder block, catching the pistons as they emerge

7.3a Prise out and discard the circlips

7.3b Displace gudgeon pin to release the pistons

7.5a Chain guide is retained by two bolts

7.5b Chain tensioner is held by four bolts

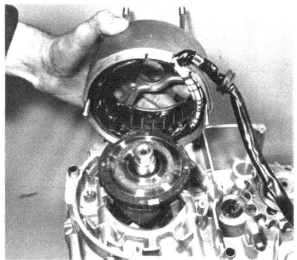

8.1 Stator is retained by two screws, note locating pin

8.2 Be **extremely** careful when using puller on rotor

9.1 Chain can be used to facilitate nut removal

10.1a Filter housing cover is retained by two screws

10.1b Remove central bolt to free filter element

11.2 Remove the clutch bolts and springs

11.3 Withdraw complete clutch plate assembly

11.5a Lock clutch centre and remove nut and washers

11.5b Remove the clutch centre, followed by ...

11.5c ... washers and thrust bearing. Remove clutch drum

12 Dismantling the engine/gearbox: removing the oil pump

1   The oil pump is retained by three screws to the inside of the right-hand outer casing. In order to gain access to the retaining screws, it is first necessary to remove the tachometer drive assembly, and the oil pump drive pinion.
2   The tachometer drive assembly can be withdrawn from the bore in the casing after slackening the clamp bolt which retains it. Place it to one side to await reassembly. Slacken and remove the oil pump pinion securing nut, and draw the tachometer driving worm off the spindle. Note that it shares a common woodruff key with the oil pump drive pinion. The oil pump drive pinion can now be removed. Should it prove stubborn, it can be eased off by levering it upwards with a pair of screwdrivers. Take great care not to damage the pinion or pump body during removal.
3   With the tachometer drive and pump pinion removed, the oil pump retaining screws can be removed, and the oil pump body lifted off. Note that two locating dowels protrude from the mounting face in the cover to ensure correct alignment of the pump assembly.

13 Dismantling the engine/gearbox: removing the gearchange mechanism, crankshaft pinion and kickstart mechanism

1   Remove the chain guide from the left-hand side of the unit. It is retained by two bolts, and can be slid off the crossover shaft when these have been removed. At the base of the protuding end of the shaft is an 'E' clip which locates in a groove in the shaft. This should be removed.
2   Turning attention to the right-hand side of the unit, disengage the selector claws from the end of the selector drum and pull the selector mechanism and crossover shaft out of the unit. Remove the two bolts which retain the stopper mechanism spring anchor plate, followed by the large diameter shouldered bolt which forms the stopper pivot. The stopper mechanism can be disengaged from the selector drum and removed as an assembly.
3   Pass a bar through one of the small end eyes and support each end on wooden blocks, to prevent the crankshaft assembly from turning. Slacken the mainshaft pinion retaining nut, and

remove it, together with the oil pump drive pinion and the mainshaft pinion.
4   Disengage the kickstart return spring from its locating pin, and release the spring tension by allowing the spring to unwind slowly. The kickstart assembly can now be withdrawn from the case. Note that the assembly can be removed by dismantling each component from the shaft if desired, though nothing is to be gained by doing this.
5   Remove the circlip and retainer from the starter driven pinion The two retaining ring halves can now be displaced, and the pinion removed, followed by the inner pinion and its friction clip. The small driving pinion can be left in position at this stage.

14 Dismantling the engine/gearbox: separating the crankcase halves

1   Having removed the various components described in Sections 6 to 13, the crankcases are ready for separation. If it is still in position, release the camshaft drive chain and draw it out of the crankcase. The crankcase halves are retained by eighteen nuts and studs which pass through the lower casing half.
2   Each of these nuts is identified with a number cast into the casing, immediately adjacent to it. This serves to indicate which nuts are holding the crankcase halves, and in which order they are to be slackened. Start with number 18 and work progressively through to number 1, slackening each one slightly, to avoid any risk of warpage. Note that there are four additional securing bolts which pass down through the upper casing half. These are to be found at the rear of the upper crankcase half grouped around the oil filler neck.
3 With the aforementioned nuts and bolts removed, the upper crankcase half can be drawn off the lower half. If separation should prove difficult, check that none of the retaining nuts or bolts has been missed. If necessary, a soft-faced mallet may be used to jar the casings apart, but take great care not to damage any of the more fragile parts, such as the cast extension which retains the right-hand outer cover.
4   As the halves draw apart, all components, with the exception of the gearbox selector drum and forks, should remain in the lower casing half. Catch the connecting rods as they emerge from the crankcase mouths, to prevent them dropping onto the gearbox internals.

1 Crankcase assembly
2 Dowel pin - 2 off
3 Dowel pin
4 Stud - 2 off
5 Stud - 4 off
6 Stud
7 Stud - 3 off
8 Washer - 8 off
9 Nut - 8 off
10 Stud
11 Stud - 2 off
12 Stud
13 Sealing washer - 6 off
14 Domed nut - 6 off
15 Bolt - 2 off
16 Bolt
17 Bolt
18 Washer - 4 off
19 Sealing washer
20 Filler plug/dipstick
21 Sealing washer
22 Drain plug
23 Sealing washer

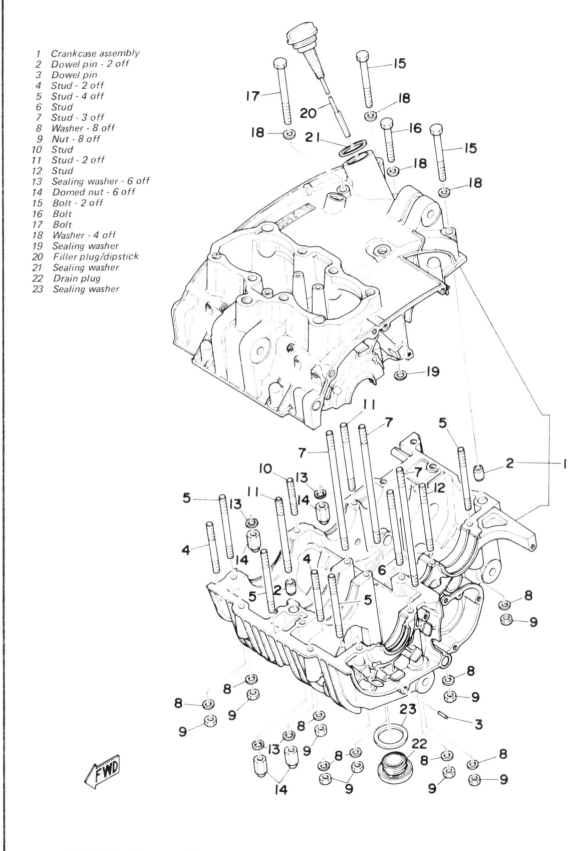

Fig. 1.2. Crankcase assembly

12.2a Slacken clamp bolt and withdraw tachometer drive

12.2b Remove nut, worm and oil pump pinion

12.3 Oil pump is retained by three bolts, and located by dowels

13.1a Detach the drive chain guide

13.1b Prise the 'E' clip from the selector shaft

13.2a Disengage the selector claw, and withdraw the shaft

13.2b Release the stopper mechanism, and ...

13.2c ... spring anchor plate

13.3a Lock crankshaft and remove nut, oil pump drive pinion and ...

13.3b ... crankshaft pinion

13.4 Release kickstart spring and withdraw assembly

13.5a Remove circlip and retaining cap, to release ...

13.5b ... retainer halves. Remove the pinion assembly

14.2a Casing bolts and nuts are numbered (see arrows)

14.2b Remove the nuts and bolts in numerical sequence

14.2c Do not omit this nut

14.3 Crankcase halves can now be separated

15 Dismantling the engine/gearbox: removing the crankcase components

1   The lower crankcase half components can be lifted out after separation, and placed to one side to await further attention. Make a note of the disposition of the half-rings which locate the crankshaft and gearshafts in relation to the crankcase.

2   The gear selector drum, together with the selector forks and shaft, are contained in the upper crankcase half. The selector fork shaft is retained axially by the end of the stopper spring anchor plate, which engages in a slot in the end of the shaft. As the stopper assembly will have been removed at an earlier stage, the shaft can be displaced and drawn out of its bore.

3   Remove the large hexagon-headed neutral detent plunger assembly from the top of the casing. This comprises a hollow bolt, sealing washer, spring and plunger. Remove also the neutral indicator switch from the top of the casing, to obviate any risk of damage during the removal of the selector drum.

4   Each of the three selector forks, which are arranged concentrically to the selector drum, has a cam follower pin and roller which engages in the selector drum track. The follower pins are each retained by a split pin. Pull out the split pins and shake

out the cam follower pin and roller from its bore in the fork. It may be found difficult to gain access to the split pins, in which case it is permissible to use a cold chisel to remove the heads, and then drive out the remainder of the pins, using a small parallel punch.

5 The three selector forks are now free to float on the selector drum. The latter should be drawn carefully out of the casing, noting the order and disposition of each fork as it falls free. It is a wise precaution to mark each fork with an arrow denoting front and a number denoting position, to aid reassembly.

16 Examination and renovation : general

1 Before examining the component parts of the dismantled

engine/gear unit for wear, it is essential that they should be cleaned thoroughly. Use a paraffin/petrol mix to remove all traces of oil and sludge which may have accumulated within the engine.

2 Examine the crankcase castings for cracks or other signs of damage. If a crack is discovered, it will require professional attention or in an extreme case, renewal of the casting.

3 Examine carefully each part to determine the extent of wear. If in doubt, check with the tolerance figures wherever they are quoted in the text. The following sections will indicate what type of wear can be expected and in many cases, the acceptable limits.

4 Use clean, lint-free rags for cleaning and drying the various components, otherwise there is risk of small particles obstructing the internal oilways.

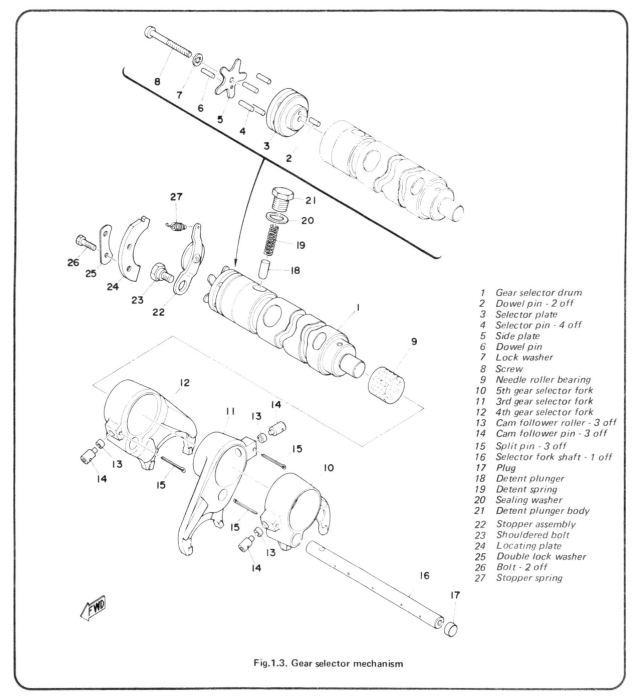

1   Gear selector drum
2   Dowel pin - 2 off
3   Selector plate
4   Selector pin - 4 off
5   Side plate
6   Dowel pin
7   Lock washer
8   Screw
9   Needle roller bearing
10 5th gear selector fork
11 3rd gear selector fork
12 4th gear selector fork
13 Cam follower roller - 3 off
14 Cam follower pin - 3 off
15 Split pin - 3 off
16 Selector fork shaft - 1 off
17 Plug
18 Detent plunger
19 Detent spring
20 Sealing washer
21 Detent plunger body
22 Stopper assembly
23 Shouldered bolt
24 Locating plate
25 Double lock washer
26 Bolt - 2 off
27 Stopper spring

Fig.1.3. Gear selector mechanism

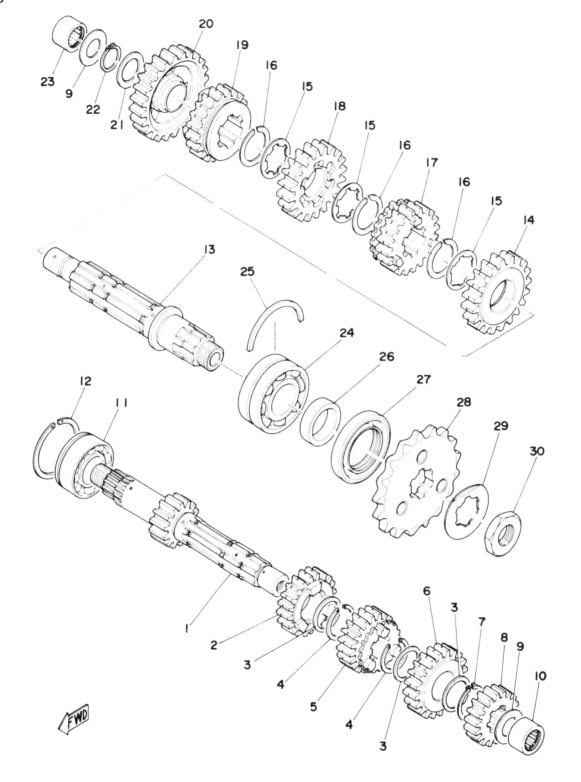

Fig.1.4. Gearbox components

1	Input shaft	9	Thrust washer - 2 off
2	4th gear pinion (21T)	10	Needle roller bearing
3	Washer - 3 off	11	Journal ball bearing
4	Circlip - 2 off	12	Circlip
5	3rd gear pinion (20T)	13	Output shaft
6	5th gear pinion (23T)	14	2nd gear pinion (27T)
7	Circlip	15	Thrust washer - 3 off
8	2nd gear pinion (17T)	16	Circlip - 3 off

17	5th gear pinion (22T)	24	Journal ball bearing
18	3rd gear pinion (26T)	25	Circlip
19	4th gear pinion (23T)	26	Collar
20	1st gear pinion (32T)	27	Oil seal
21	Washer	28	Drive sprocket (17T)
22	Circlip	29	Lock washer
23	Needle roller bearing	30	Nut

15.1a Lift out the crankshaft assembly ...

15.1b ... followed by the ...

15.1c ... gear clusters

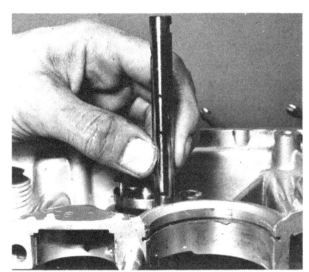

15.2 Withdraw the selector fork shaft

15.3a Remove the neutral detent plunger assembly ...

15.3b ...followed by the neutral indicator switch

15.4 Pull out the split pins to free the cam follower pins

15.5 Selector drum can be pulled out of casing and forks removed

17 Crankshaft assembly: examination and renovation

1   The crankshaft assembly can be regarded as two separate sets of flywheels and connecting rods, pressed together by means of a central coupling which carries the camshaft chain drive sprocket. It is not possible to separate the crankshaft assembly without access to the appropriate press equipment or to re-align the dismantled assembly to a sufficiently high standard of accuracy by amateur means. In consequence, it is imperative that the complete flywheel is entrusted to a Yamaha Service Agent will have either the necessary repair facilities or a service-exchange replacement.

2   Failure of the big end bearings is invariably accompanied by a whirring noise from within the crankcase which progressively gets worse. Some vibration will also be experienced. There should be no vertical play whatsoever in either of the connecting rods after the old oil has been washed out of the bearings. If even a small amount of play is evident, the bearing concerned is due for replacement. Do not run the machine with worn big end bearings, otherwise there is a risk of causing extensive damage by the breakage of a connecting rod or the crankshaft.

3   Do not confuse big end wear with side play, a certain amount of which is acceptable in the big end bearing assembly. It is permissible to move the small end of each connecting rod sideways not more than 2 mm (0.08 inches) if the amount of sideplay is within acceptable limits.

4   The crankshaft is carried on four main bearings, the outer right-hand bearing being of the journal ball type, whilst the other three are of the roller type. If wear is evident in the form of play or if the bearings feel rough as they are rotated, they should be renewed. This again is a specialist repair job, necessitating the services of a Yamaha agent. The crankshaft assembly must be separated to gain access to the two innermost bearings and re-aligned with great accuracy after these bearings have been replaced. It is essential that the correct replacement bearings are fitted and not pattern parts. The bearings are grooved to correspond with the bearing retainers and have a drilling to locate with a dowel pin in the bearing housing.

5   Failure of the main bearings is usually characterised by an audible rumble from the bottom end of the engine, accompanied by vibration. The vibration will be especially noticeable through the footrests.

18 Connecting rods: examination and renovation

1   It is unlikely that either of the connecting rods will bend

during normal usage unless an unusual occurrence such as a dropped valve has caused the engine to lock. Carelessness when removing a tight gudgeon pin can also give a rise to a similar problem. It is not advisable to straighten a bent connecting rod; renewal is the only satisfactory solution.

2   The small end eye of the connecting rod is unbushed and it will be necessary to renew the connecting rod if the gudgeon pin becomes a slack fit. Always check that the oil hole in the small end eye is not blocked since if the oil supply is cut off, the bearing surfaces will wear very rapidly.

19 Cylinder block: examination and renovation

1   The usual indications of badly worn cylinder bores and pistons are excessive oil consumption accompanied by blue smoke from the exhausts and pistons slap, a metallic rattle that occurs when there is little or no load on the engine. If the top of each cylinder bore is examined carefully, it will be found there is a ridge on the thrust side, denoting the limit of travel of the uppermost piston ring. The depth of this ridge will vary according to the amount of wear that has taken place.

2   Measure the bore diameter just below the ridge, using an internal micrometer. Compare this reading with the diameter close to the bottom of the cylinder bore, which has not been subjected to wear. If the difference in readings exceeds 0.005 inch (0.127 mm) it is necessary to have the cylinder block rebored and to fit oversize pistons and rings.

3   If an internal micrometer is not available, the amount of wear can be checked by inserting each piston in turn (without rings) into the bore with which it was previously associated. If it is possible to insert a 0.004 inch (0.101 mm) feeler gauge between the piston and the cylinder wall on the thrust side, remedial action must be taken.

4   Check the surface of each cylinder bore for score marks or other damage that may have resulted from an earlier engine seizure or displacement of a gudgeon pin and/or circlip. A rebore will be necessary to remove any deep indentations, irrespective of the amount of bore wear, otherwise a compression leak will occur.

5   Check that the external cooling fins are not clogged with oil or road dirt, otherwise the engine will overheat. The fins can be cleaned with a wire brush, provided care is taken to ensure the fins are not broken or badly scratched.

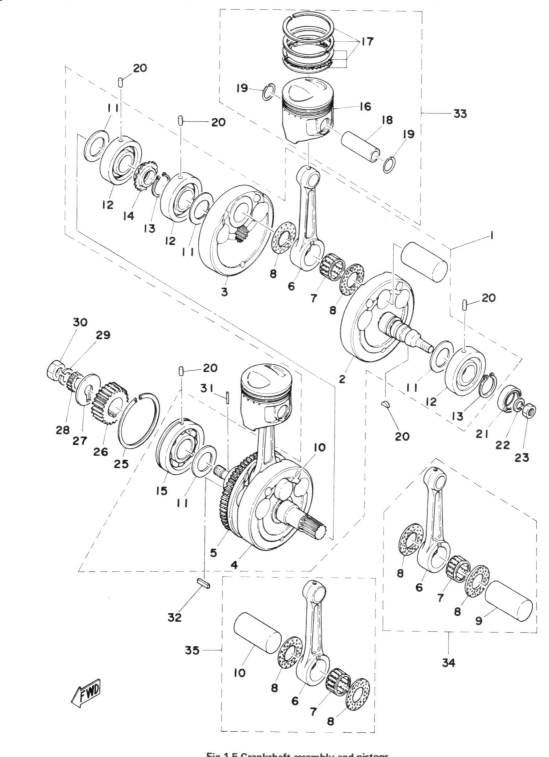

Fig.1.5 Crankshaft assembly and pistons

1   Crankshaft assembly complete
2   Flywheel left-hand cylinder
3   Flywheel, left-hand cylinder
4   Flywheel, right-hand cylinder
5   Flywheel, right-hand cylinder
6   Connecting rod - 2 off
7   Caged roller bearing - 2 off
8   Thrust washer - 4 off
9   Crankpin
10 Crankpin
11 Shim - 4 off
12 Roller bearing - 3 off

13 Circlip - 2 off
14 Camshaft drive sprocket
15 Journal ball bearing
16 Piston - 2 off
17 Piston ring set - 2 off
18 Gudgeon pin - 2 off
19 Circlip - 4 off
20 Dowel pin - 4 off
21 Oil seal
22 Spring washer
23 Nut
24 Woodruff key

25 Circlip
26 Primary drive pinion
27 Lock washer
28 Oil pump drive gear
29 Spring washer
30 Nut
31 Dowel pin
32 Woodruff key
33 Piston assembly
34 Connecting rod assembly (left-hand)
35 Connecting rod assembly (right hand)

20 Pistons and piston rings - examination and renovation

1   If a rebore is necessary, the existing pistons and rings can be disregarded as they will be replaced with their oversize equivalents as a matter of course.

2   Remove all traces of carbon from each piston crown, using a soft scraper to ensure the surface is not marked. Finish off by polishing each crown with metal polish, to give a smooth mirror-like surface to which carbon will not adhere so readily in the future. Never use emery cloth, the particles of which will embed in the soft aluminium alloy.

3   Piston wear usually occurs at the skirt or lower end of the pistons and takes the form of vertical streaks or score marks on the thrust side. There may also be some variation in the thickness of the skirt. Light scoring may be polished out using fine emery cloth, although where this has been caused by the piston 'picking up' (localised seizure) the surface will probably have become too badly damaged and work-hardened for any renovation to be practicable.

4   The piston ring grooves may have become enlarged in use, allowing excessive clearance, and consequently, loss of compression. With the rings in position on their respective pistons, measure each one in turn by trying feeler gauges in the gap between the ring and its groove. The nominal clearance should be 0.04 − 0.08 mm (0.0016 − 0.0032 in). If the gap should be in excess of 0.15 mm (0.006 in), the piston(s) should be renewed. Note that the above information applies to the top two rings only. The oil ring, which is of the expander type, should have no clearance. If any clearance is evident, it indicates that the expander has become fatigued, in which case the oil ring assembly must be renewed.

5   Piston ring wear can be assessed by inserting each ring in turn into its respective cylinder bore. Use the bottom of the piston to push the ring squarely into position, about one inch below the top of the cylinder. Measure the end gap of each ring with feeler gauges, noting that, in the case of the oil ring, the expander should be omitted. The gaps should be as follows:

	Nominal end-gap	Wear limit
Compression ring (top)	0.2-0.4 mm (0.008-0.016 in)	0.8 (0.031 in)
Wiper ring (middle)	0.2-0.4 mm (0.008-0.016 in)	0.8 (0.031 in)
Oil scraper ring (bottom)	0.3-0.6 mm (0.012-0.024 in)	1.0 (0.039 in)

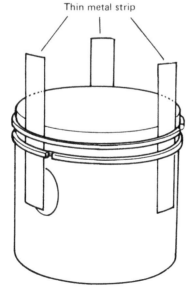

Thin metal strip

Fig.1.6. Removing piston rings

21 Piston and piston rings: identification marks

1   The piston, piston rings and cylinder bores are each marked with various code letters to enable each component to be correctly matched with the others. The purpose of this Section is to enable the owner to identify the parts required to match existing components, an important point if, for example, a second hand cylinder block is obtained.

2   The piston crown will be found to be marked with an arrow, indicating front, and a number. The nominal size of the piston is 75 mm, the number on the piston crown giving the exact diameter of that component. For example, if the number stamped on the crown is 0.956, it indicates that the actual diameter is 74.956 mm, or 0.046 mm below the nominal size.

3   The actual bore size of a standard cylinder is stamped at the bottom of the related bore. For example, if the number found is .007, the actual bore size will be 75.007 mm. The piston to bore clearance can be calculated from the two above examples. In this case it would be 0.053 mm clearance.

4   The compression (top) and wiper (middle) rings are marked with an R, which must face upwards when installed on the piston. On oversize rings, the ring size is found opposite the R mark on the other side of the ring gap. Thus a ring marked R50 indicates that it is suited to a 2nd oversize, or + 0.50 mm piston.

5   The individual rails which comprise the oil scraper ring (bottom) can be fitted either way up on the piston. They are, however colour-coded to indicate their size, by way of one or two painted strips opposite the end gap. The coding is as follows:

Colour	Indicated size
Blue (one mark)	Standard
Blue (two marks)	1st Oversize (+0.25 mm)
Red (one mark)	2nd Oversize (+0.50 mm)
Red (two marks)	3rd Oversize (+0.75 mm)
Yellow (One mark)	4th Oversize (+1.00 mm)

21.2 Note arrow denoting front, and size markings

22 Cylinder head and valves: examination and renovation

1   It is best to remove all carbon deposits from the combustion chambers before removing the valves for inspection and grinding-in. Use a blunt end chisel or scraper so that the surfaces are not damaged. Finish off with a metal polish to achieve a smooth, shining surface. If a mirror finish is required a high speed felt mop and polishing soap may be used. A chuck attached to a flexible drive will facilitate the polishing operation.

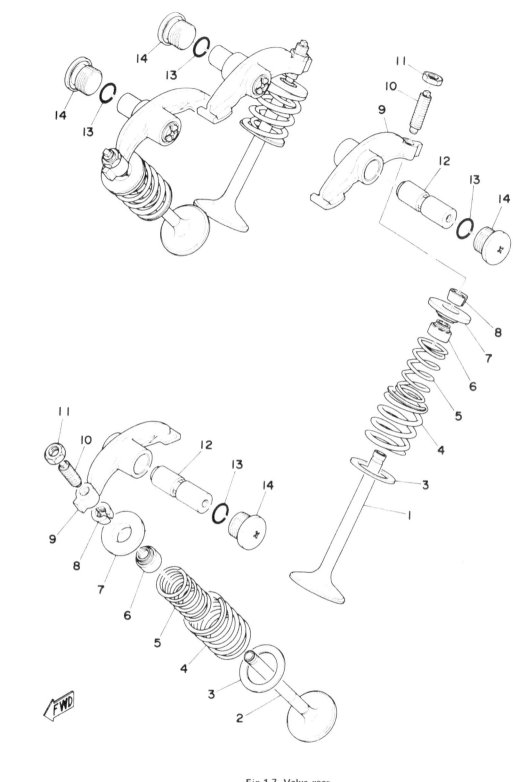

Fig.1.7. Valve gear

1	Inlet valve - 2 off	8	Split collets - 8 off
2	Exhaust valve - 2 off	9	Valve rocker arm - 4 off
3	Valve spring seat - 4 off	10	Adjusting screw - 4 off
4	Outer valve spring - 4 off	11	Lock nut - 4 off
5	Inner valve spring - 4 off	12	Rocker shaft - 4 off
6	Valve stem oil seal - 4 off	13	'O' ring - 4off
7	Valve spring seat - 4 off	14	Plug - 4 off

2　A valve spring compression tool must be used to compress each set of valve springs in turn, thereby allowing the split collets to be removed from the valve cap and the valve springs and caps to be freed. Keep each set of parts separate and mark each valve so that it can be replaced in the correct combustion chamber. There is no danger of inadvertently replacing an inlet valve in an exhaust position, or vice-versa, as the valve heads are of different size. The normal method of marking valves for later identification is by centre punching them on the valve head. This method is not recommended on valves, or any other highly stressed components, as it will produce high stress points and may lead to early failure. Tie-on labels, suitably inscribed, are ideal for this purpose.

3　Before giving the valves and valve seats further attention, check the clearance between each valve stem and the guide in which it operates. Clearances are as follows:

	Standard	Wear limit
Inlet valve/guide clearance		
	0.020 – 0.044 mm	0.100 mm
	(0.0008–0.0017 in)	(0.0039)
Exhaust valve/guide clearance		
	0.035 – 0.059 mm	0.120 mm
	(0.0014–0.0023 in)	(0.0047 in)

Measure the valve stem diameter at the point of greatest wear and then measure again at right-angles to the first measurement. If the valve stem diameter is below the service limit it must be renewed.

Inlet valve stem diameter
　　7.975–7.99 mm　　　(0.3140–0.3146 in)

Exhaust valve stem diameter
　　7.96–7.975 mm　　　(0.3134–0.3140 in)

The valve stem/guide clearance can be measured with the use of a dial guage and a new valve. Place the new valve into the guide and measure the amount of shake with the dial gauge tip resting against the top of the stem. If the amount of wear is greater than the wear limit, the guide must be renewed.

4　To remove an old valve guide, place the cylinder head in an oven and heat it to about 150ºC. Care should be taken to ensure there is no risk of distortion. If unfamiliar with this task, seek professional advice first. The old guide can now be tapped out

from the cylinder side. The correct drift should be shouldered with the smaller diameter the same size as the valve stem and the larger diameter slightly smaller than the O/D of the valve guide removal, local expansion may be effected by the use of a blow torch. Do not allow the flame to remain on one spot for too long, and ensure that the head is evenly heated. Each valve guide is fitted with an 'O' ring to ensure perfect sealing. The 'O' rings must be replaced with new components. New guides should be fitted with the head at the same heat as for removal.

23 Rocker arms and spindles: removal, examination and renovation

1　The valve rocker arms are carried on spindles in the cylinder head cover, these being retained by sleeves in the mounting hole bores. Push out the headed sleeves, noting the 'O' rings which form an oil seal between the cylinder head cover and each sleeve. Remove the chromium plated end caps and 'O' rings from the spindle bores. The spindles can be removed by screwing a 6 mm bolt into the thread in each spindle end, and drawing the spindle out. The rockers will drop free as the spindles are removed.

2　Examine the rocker arm rubbing faces for signs of wear or scoring. If any great degree of damage is evident it is likely that the camshaft will be correspondingly scored, in which case both components will be in need of renewal.

3　Check the rocker arm bores for wear, together with the spindles on which they pivot. It is unlikely that these will wear excessively unless the oil feed to them has at some time become obstructed. Measure the rocker arm bore, and its associated spindle diameter. Compare the readings with those given in the following table:

Rocker arm bore diameter:	15.01 mm(0.5909in) nominal
Rocker spindle diameter:	14.98 mm(0.5898in) nominal
Rocker arm to spindle clearance:	0.05 mm(0.002in) nominal
Rocker arm to spindle wear limit:	0.10 mm(0.004in)

4　When reassembling the rocker arms and spindles, ensure that they are clean and well lubricated. Note that the threaded hole in the spindle must face outwards. If this precaution is not observed, it will prove most difficult to remove the spindles on subsequent occasions. Ensure that the retaining sleeve 'O' rings are in good condition, to preserve oil-tightness.

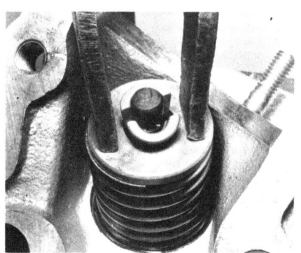

22.2a Compress spring to release collets

22.2b Remove upper spring seat, ...

22.2c ... followed by valve springs

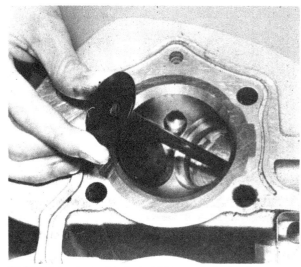

22.2d Valve can now be displaced from cylinder head

22.4 Note oil seal on top of valve guide

23.1a Remove spindle end caps ...

23.1b ... and dislodge sleve and 'O' ring

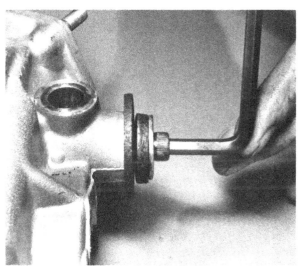

23.1c Use 6mm bolt to draw out rocker spindle

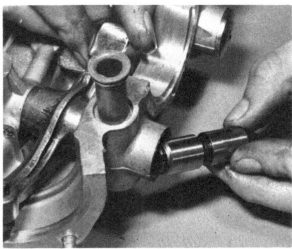

23.4 Thread in spindle must face outwards on reassembly

24 Camshaft, drive chain and tensioner: examination and renovation

1   The single overhead camshaft used on all the Yamaha 650 twins is carried on two pairs of single-row ball races, and takes its drive from the centrally-mounted crankshaft drive sprocket, via an endless chain. The drive and driven sprocket give a reduction ratio of 2:1, effectively turning the camshaft at half engine speed.
2   Wear in the camshaft drive chain is characterised by a persistent rattling which will prove impossible to eradicate by adjusting the tensioner. The normal cause of chain wear is over or under tightening of the tensioner, or sheer old age. It should be noted that the camshaft and drive chain can be removed for examination after the cylinder head cover has been removed. If it is not wished to dismantle the engine further, a length of wire, or better still an old chain, should be attached to one end of the chain before it is removed, to act as a guide during installation.
3   The chain should be carefully washed in petrol and dried prior to examination. Bend the chain from side to side, noting any signs of sloppiness. Look also for tight spots in the chain. Unless the chain is in obvious good health it should be renewed, as the need for subsequent renewal, where its poor condition has not been realised, is both time consuming and annoying. If in doubt as to the serviceability or otherwise of the chain, seek the advice of a Yamaha Service Agent, who will be able to advise if a new chain is required.
4   Examine the crankshaft and camshaft sprockets for wear or damage. Chipped or missing teeth will be immediately obvious, and will necessitate renewal of the component concerned. Wear, in the form of hooked teeth, may also warrant renewal. In this latter case, however, consult a Yamaha Service Agent first, as renewal of either sprocket will be costly, and in the case of the crankshaft sprocket, will involve a complete engine stripdown.
5   The general condition of the camshaft should be checked, looking for signs of scoring or abrasion on the rubbing faces of the lobes. It is unlikely to amount to very much, unless the lubrication feed has failed at some time, in which case the camshaft will have to be renewed. It is possible to have a damaged or worn camshaft built up and re-profiled, but it should be noted that this is likely to be a costly process. It is worth remembering as a last resort, however, should a replacement be unavailable.
6   Four ball races support the camshaft, being arranged in pairs at each end of the shaft. The bearings should be washed out in petrol and dried. Spin each bearing to test for roughness. Should this or any other sign of wear be present, renew the bearings, preferably as a set.

7   Check the condition of the chain tensioner assembly, and the chainguide, for wear or damage. Should the rubbing surfaces of either part appear badly worn, the component should be renewed.

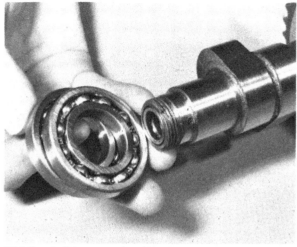

24.6 Camshaft has two pairs of bearings

25 Clutch assembly: examination and renovation

1   After an extended period of service the clutch linings will wear and promote clutch slip. The limit of wear measured across each inserted plate and the standard measurement is as follows:

	Standard	Service limit
Clutch plate thickness	3.5 mm (0.140 in)	3.1 mm (0.122 in)

When the overall width reaches the limit, the inserted plates must be renewed, preferably as a complete set.
2   The plain plates should not show any excess heating (blueing). Check the warpage of each plate using plate glass or surface plate and a feeler guage. The maximum allowable warpage is 0.2 mm (0.008 in).
3   Check the condition of the push rods by rolling them on a surface plate. Review if out of true or scored badly.
4   Check the free length of each clutch spring with a vernier gauge. After considerable use the springs will take a permanent set thereby reducing the pressure applied to the clutch plates. The correct measurements are as follows:

	Standard	Service limit
Clutch springs	3.46 mm (0.136 in)	2.46 (0.097 in)

5   Examine the clutch assembly for burrs or indentations on the edges of the protruding tongues of the inserted plates and/or slots worn in the edges of the outer drum with which they engage. Similar wear can occur between the inner tongues of the plain clutch plates and the slots in the clutch inner drum. Wear of this nature will cause clutch drag and slow disengagement during gear changes, since the plates will become trapped and will not free fully when the clutch is withdrawn. A small amount of wear can be corrected by dressing with a fine file; more extensive wear will necessitate renewal of the worn parts.
6   The clutch release mechanism attached to the final drive sprocket cover does not normally require attention provided it is greased at regular intervals. It is held to the cover by two cross-heads screws and operates on the worm and quick start thread principle.
7   The thrust bearing should be cleaned, checked for wear and greased, prior to reassembly. It does not normally wear significantly in use.

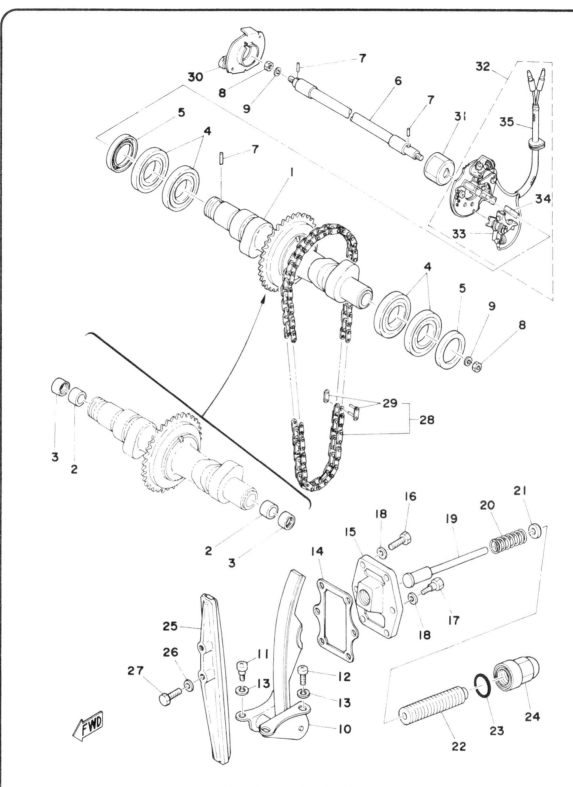

Fig.1.8. Camshaft, camshaft drive and tensioners

1 Camshaft assembly	10 Tensioner bracket	19 Tensioner rod	28 Camshaft chain
2 Bearing - 2 off	11 Dowel screw - 2 off	20 Spring	29 Joining link
3 Oil seal - 2 off	12 Screw - 2 off	21 Spring seat	30 Automatic timing unit (ATU)
4 Bearing - 4 off	13 Spring washer - 4 off	22 Adjuster	31 Contact breaker cam
5 Oil seal - 2 off	14 Gasket	23 'O' ring	32 Contact breaker and base plate
6 ATU shaft	15 Tensioner body	24 Adjuster cap	assembly
7 Dowel pin - 3 off	16 Bolt - 4 off	25 Chain guide	33 Contact breaker assembly-2 off
8 Nut - 2 off	17 Dowel bolt -2 off	26 Sealing washer - 2 off	34 Lubricating wick
9 Spring washer - 2 off	18 Plain washer - 2 off	27 Bolt - 2 off	35 Low tension leads

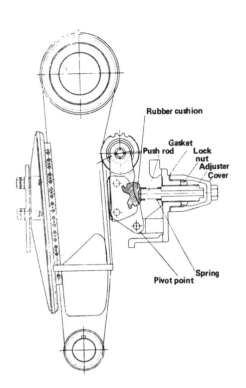

1 Clutch assembly
2 Clutch outer drum & pinion (72T)
3 Bearing
4 Thrust plate
5 Clutch centre
6 Friction plate - 7 off
7 Clutch plate - 6 off
8 Pressure plate
9 Spring - 6 off
10 Screw - 6 off
11 Pushrod
12 Nut
13 Belled washer
14 Plain washer
15 Spacer
16 Thrust plate
17 Thrust washer
18 Ball bearing - 2 off
19 Push rod
20 Push rod
21 Push rod seal
22 Quick start worm
23 Dust seal
24 Bolt
25 Nut
26 Quick start worm housing
27 Screw - 2 off
28 Trunnion
29 Clevis pin
30 Split pin
31 Spring
32 Spring hook

Fig.1.9. Early type camshaft chain tensioner

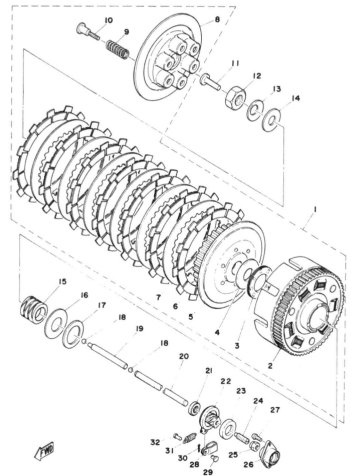

Fig. 1.10. Clutch assembly – TX650 and all XS650 models

26 Gearbox components - examination and renovation

1 Examine each of the gear pinions to ensure that there are no chipped or broken teeth and that the dogs on the end of the pinions are not rounded. Gear pinions with any of these defects should be removed from their shafts and replaced by new gears.

2 The gearbox bearings must be free from play and show no signs of roughness or jamming when rotated. Once again it is necessary to make sure any old oil is removed from the bearings before they are inspected. The gearbox incorporates both caged needle roller and journal ball bearings which can be tapped out for renewal once the oil seals have been prised out of place.

3 It is advisable to renew the gearbox oil seals irrespective of their condition. Should a re-used oil seal fail at a later date, a considerable mount of dismantling is necessary to gain access and renew it.

4 Check the gear selector rods for straightness by rolling them on a sheet of plate glass. A bent rod will cause difficulty in selecting gears and will make the gearchange action particularly heavy. The remaining circlips on the fork shafts must be removed in order to rest them. Replace the circlips once the test is complete.

5 The selector forks should be examined closely, to ensure that they are not bent or badly worn. Wear is unlikely to occur unless the gearbox has been run for a period with a particularly low oil content.

6 The tracks in the gear selector drum, with which the selector forks engage, should not show any undue signs of wear unless neglect has led to under-lubrication of the gearbox. Check that the plunger spring bearing on the cam plate plunger has not lost its action and that the springs of the gearchange lever pawl assembly have good tension. Any damage to, or weakness of, the gearchange lever return spring will be self-evident.

7 If the kickstart has shown a tendency to slip , it will be necessary to examine the kickstart gear and shaft. The kickstart assembly can be inspected without dismantling, the most likely cause of failure being a broken kickstart spring clip. If it is necessary to renew any part of the kickstart assembly proceed as follows:

8 Remove the circlip nearest the shaft's splined end, and slide off the spring cover, spring and spring guide. Remove the large diameter circlip that encircles and retains the kickstart gear holder which is in the form of two collets. The spacing shim and kickstart gear can now be removed. If the kickstart spring has any signs of weakness or stress deformity it should be renewed.

26.1a 4th gear pinion is retained by washer and clip to the input shaft

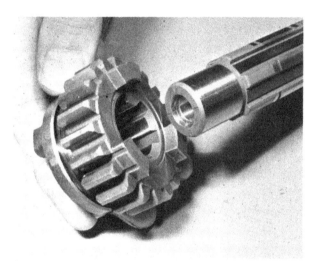

26.1b Fit the 3rd gear pinion, ...

26.1c ... circlip and washer

26.1d The 5th gear pinion is fitted next, ...

26.1e ... followed by its washer and circlip

26.1f Finally, fit the 2nd gear pinion

26.1g Needle roller bearing is preceded by thrust washer

26.1h Bearing with locating clip is fitted to other end of shaft

26.1i Slide 2nd gear pinion onto the output shaft, ...

26.1j ... followed by 5th gear pinion. Note ...

26.1k ... circlip and washer either side

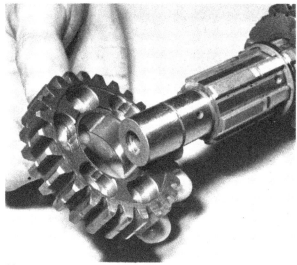

26.1l Fit the 3rd gear pinion ...

26.1m...and retain with washer and circlip

26.1n 4th gear pinion is fitted next, ...

26.1o ... followed by 1st gear pinion

26.1p Shaft is supported by needle roller bearings

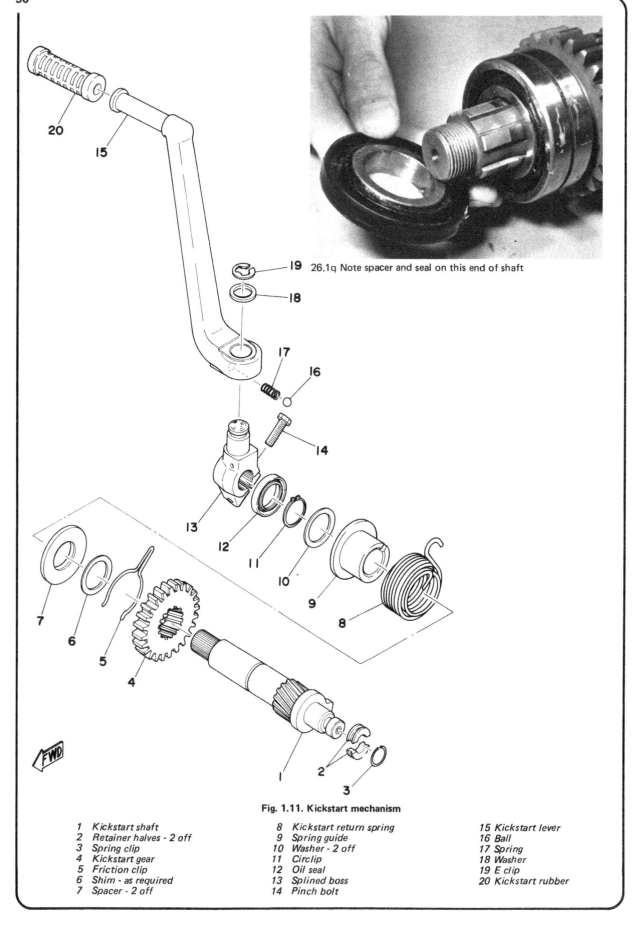

26.1q Note spacer and seal on this end of shaft

Fig. 1.11. Kickstart mechanism

1 Kickstart shaft
2 Retainer halves - 2 off
3 Spring clip
4 Kickstart gear
5 Friction clip
6 Shim - as required
7 Spacer - 2 off

8 Kickstart return spring
9 Spring guide
10 Washer - 2 off
11 Circlip
12 Oil seal
13 Splined boss
14 Pinch bolt

15 Kickstart lever
16 Ball
17 Spring
18 Washer
19 E clip
20 Kickstart rubber

27 Oil pump and tachometer drive: examination and renovation

1 With the oil pump removed as described in Section 12 of
this Chapter, the pump components should be examined for
wear. The lobe faces must be free of scoring, (normally caused
by contaminated oil), as should the machined side faces of the
pump body and mating surface. Assemble the inner and outer
rotor in the pump body and ensure that the pump turns freely,
but without excessive play. On reassembly, ensure that both
reference marks (one on each rotor) face outwards and that the
assembly is well lubricated.
2 The tachometer drive gears and spindle should be greased on
reassembly. Little else can be done by way of maintenance.
If excessively worn, renew these components.

27.1 Examine the oil pump components for wear and scoring

28 Engine reassembly: replacing the gear selector drum and forks

1 Arrange the three selector forks in the upper casing half,
ensuring that each fork is in its correct position. Slide the pivot
shaft into position to retain the selector forks, noting that the
slotted end of the pivot shaft must be on the right-hand side of
the unit.
2 The selector drum can now be fed into position, ensuring that
the small diameter journal is correctly engaged with its needle
roller bearing. Fit the roller to each of the cam follower pins,
having first greased the pins to retain their respective rollers,
as they are fitted to the selector forks bores. Feed each follower
pin assembly in turn, turning the selector fork until it drops
into its operating track. Ensure that the holes in each pin align
with the retaining holes in the selector forks.
3 Use new split pins to retain the follower pins in position. In
the case of the outer forks, make sure when fitting the split pins
that they are such a position that subsequent removal will be
possible. Check the operation of the selector drum, having first
liberally lubricated the assembly with clean engine oil.

29 Engine reassembly: fitting the gearbox components and crankshaft assembly

1 Lay the gearbox input and output shaft assemblies in
position in the upper casing half, ensuring that their retaining
half-rings engage correctly in the grooves in the casing. Fit the
pushrod oil seal in position in the casing. It is worth checking
at this stage that all the gears select correctly, to avoid any risk
of having to dismantle the unit again later.
2 Lower the crankshaft into position, feeding the connecting
rods through the crankcase mouths. The main bearings are
fitted with locating pins which prevent the outer races from
turning during use. Scribe lines will be found on the outer face
of the bearing as an aid to accurate alignment.
3 Fit the camshaft drive chain in position round the crankshaft
sprocket, securing the chain ends with wire to prevent the chain
from becoming displaced. Make sure that all the crankcase
components, especially the big end and main bearings, are well
lubricated with engine oil. This is important as the lubrication
system in the rebuilt unit may take a few seconds to circulate
oil when the initial start-up is made.

28.1 Fit the selector forks and slide shaft into position

28.2a Slide selector drum into position ensuring that ...

28.2b ... drum end engages with needle roller bearing

28.2c Assemble rollers on cam follower pins

28.2d Rollers must seat in tracks thus

28.2e Fit assembled rollers ensuring that holes align

28.3a Make sure that new split pins are fitted ...

28.3b ... and ends secured properly

28.3c Lubricate assembly with clean engine oil and check operation

29.1a Fit input shaft into upper casing half ...

29.1b ... followed by output shaft. Check gear selection

29.1c Ensure that pushrod seal seats correctly

29.2a Lower crankshaft assembly into position

29.2b Note locating pips for main bearings

29.2c Make sure that half rings are positioned

29.2d Scribe line should align with casing face

29.3a Camshaft chain should be positioned around crankshatt

29.3b Lubricate main and big-end bearings thoroughly

30 Engine reassembly: joining the crankcase halves

1   Ensure that the mating surfaces of both crankcase halves are
clean and oil-free, then apply a jointing compound to one face,
Note that a rubber blanking plug is fitted to a recess in the upper
crankcase half, adjacent to the crankshaft (see photograph). This
should be stuck in position with jointing compound before the
lower crankcase half is fitted.

2   Slide the lower crankcase half over the mounting studs.
Make sure that the various half rings locate correctly in their
grooves — any reluctance for the crankcase halves to seat properly
can usually be attributed to a displaced half ring. Finally, fit
the retaining nuts and bolts, tightening them evenly in the
sequence given on the lower case. Start with No 1 and work
through to No 18, followed by the upper casing half bolts.
Tighten the nuts down to 2m - kgs (14 ft - lbs).

3   If not already in position, fit the sump plate to the bottom
of the crankcase, using a new gasket. The neutral detent assembly
and neutral indicator switch can also be fitted in their respective
positions on the upper casing half.

30.1 Ensure blanking plug is refitted using gasket cement

30.2 Lower bottom half of casing into position

30.3 Fit sump plate assembly using new gasket

31 Engine reassembly: fitting the crankshaft pinion, gear selector mechanism and kickstart mechanism

1 Fit the large Woodruff key to the crankshaft end, and slide the crankshaft pinion into position. Next, fit the oil pump pinion driving pin, followed by the special slotted washer. Place the oil pump drive pinion in position and fit the spring washer and securing nut. Lock the crankshaft with a bar passed through one of the connecting rod eyes, and tighten the securing nut to 7-10 m-kgs (51-72 ft lbs).

2 Fit the stopper mechanism spring anchor plate in position, ensuring that it locates in the selector drum groove and that the end engages in the slot in the end of the pivot shaft. Fit the locking plate and the two retaining bolts. Do not forget to knock over the locking tabs when the bolts have been fully tightened. Position the stopper lever against the selector drum pins, and fit and tighten the shouldered retaining bolt.

3 Fit the cross-over shaft in its bore in the casing, ensuring that the selector claw and centring spring engage correctly. Turn the selector drum to obtain either second, third or fourth gear and check that the two tangs of the selector claw are equidistant from the selector drum pin. If necessary, adjust the selector claw by means of the eccentric pin on which the centring spring bears. The 'E' clip should be fitted to the cross-over shaft on the left-hand side of the unit, to prevent it becoming displaced.

4 The kickstart mechanism can now be fitted into the casing, ensuring that the tang of the friction clip fits into the recess in the casing. The return spring should be tensioned with a pair of pliers and hooked over its anchor pin.

32 Engine reassembly: fitting the clutch

1 Slide the small plain washer onto the end of the gearbox mainshaft, followed by the larger thrust washer. Fit the clutch bearing sleeve, then slide the clutch drum onto the sleeve. Fit the washer, thrust bearing and thrust washer in the order shown in the accompanying photographs, then place the clutch centre in position. The centre is retained by a plain washer, a dished washer and a nut, fitted in that order. Lock the clutch in the same manner as that used during the dismantling sequence, and tighten the securing nut to 7.5 - 8 m-kgs (54-58 ft lbs).

2 Referring to Fig 1.10 to establish the correct order and position of the two pushrods, fit the first pushrod, a ball, the second pushrod, another ball, and finally the mushroom headed pushrod. The clutch plates can be fitted now, starting and finishing with a friction plate. Do not forget to fit the cushion rings on XS1, XS1B and XS2 models. Finally fit the clutch cover, springs and screws. The latter should be screwed down tight.

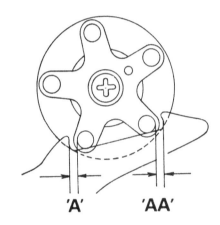

'A'   'AA'

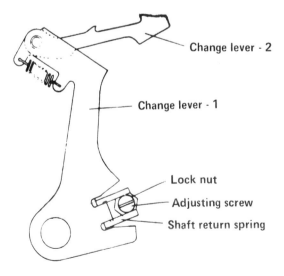

Change lever - 2

Change lever - 1

Lock nut

Adjusting screw

Shaft return spring

Fig.1.12. Gearchange claw adjustment

'A' and 'AA' must be equidistant

31.1a Fit crankshaft pinion into position, ...

31.1b ... noting driving pin for oil pump driving pinion

31.1c Fit and tighten securing nut

31.2a Replace stopper assembly and anchor plate

31.2b Fit anchor plate as shown

31.3 Slide selector shaft into casing and engage claw

32.1a Slide shim and thrust washer on to shaft ...

32.1b ... followed by bearing sleeve ...

32.1c ...and clutch outer drum

32.1d Note position of washers and thrust bearings

32.1e Fit clutch centre and securing nut

32.2a Slide pushrods, ...

32.2b ... balls and, ...

32.2c ... mushroom headed pushrod into bore

... 32.2d Assemble clutch plain and friction plates....

32.2e ... followed by clutch pressure plate

32.2f Note alignment marks on cover and centre (late models)

32.2g Fit and tighten springs and screws

32.2h Use bar through small end eye to lock crankshaft.

passing a bar through one of the connecting rod eyes then fit and tighten the securing nut to 7-7.5 m-kgs (50-54 ft. lbs).
2 Check that the locating pin is in position, then fit the alternator stator and its two retaining screws, taking care not to damage the brushes. Fit the wiring grommet in position at the bottom of the casing, and clip the cable in its guide at the rear of the unit. Depress the spring loaded washer at the top of the neutral indicator switch, and fit the cable and rubber dust shroud.

3 Make sure that the spacer is fitted correctly on the end of the mainshaft, then slide on the gearbox sprocket. Fit the tab washer in place over the splines, followed by the retaining nut. The nut has a recess on one side which must face inwards. If fitted incorrectly, the sprocket will be loose on its splines and will wear quickly. Finally, tighten the nut and knock the locking tab over. The sprocket can be locked by bunching the drive chain against the casing (see photograph). Alternatively the nut can be left slack until the unit is installed in the frame and the rear chain fitted. It can then be held by applying the rear brake. Tighten the nut to 10 - 12 m-kgs (72 - 87 ft lbs).

33 Engine reassembly: refitting the oil pump, tachometer drive and outer casing

1 Having made sure that the mating faces are clean, and that the locating dowels are in position, fit the oil pump and tighten the retaining screws. Fit the Woodruff key in its slot and slide on the drive pinion, followed by the tachometer driving worm. Fit the spring washer and securing nut.
2 Slide the tachometer drive assembly into the casing, ensuring that it meshes correctly, and tighten the clamp bolt which retains it. Clean the jointing faces of the outer casing and crankcase, and fit a new gasket. Check that the oil passages are not obstructed and that the locating dowels are in position, before placing the outer casing in position and fitting and tightening the socket screws which retain it.

34 Engine reassembly: fitting the alternator and gearbox sprocket

1 Fit the Woodruff key into its keyway in the mainshaft, and slide the alternator rotor into position. Lock the crankshaft by

33.1a Clean mating surfaces and fit locating dowels

33.1b Position oil pump. Fit screws, key and pinion

33.2a Fit worm and securing nut. Shaft is retained ...

33.2b ... by clamp bolt on outside of casing

34.1 Slide alternator rotor on to crankshaft and fit nut

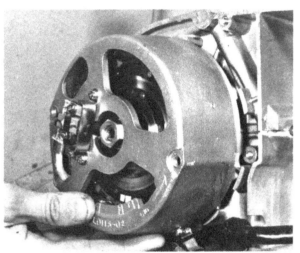

34.2 Ensure locating pin engages as stator is fitted

34.3a Recess in nut must face inwards

34.3b Lock sprocket with chain and tighten nut

35 Engine reassembly: replacing the pistons and cylinder block

1   Before replacing the pistons, place a clean rag in each crank-case mouth to prevent any displaced component from falling into the crankcase. It is only too easy to drop a circlip while it is being inserted into the piston boss, which could necessitate a further strip down for its retrieval.

2   Fit the pistons onto their original connecting rods, with the arrow embossed on each piston crown facing forwards. If the gudgeon pins are a tight fit in the piston bosses, warm each piston first to expand the metal. Do not forget to lubricate the gudgeon pin, small end eye and the piston bosses before reassembly.

3   Use new circlips, NEVER re-use old circlips. Check that each circlip has located correctly in its groove. A displaced circlip will cause severe engine damage.

4   Fit a new cylinder base gasket over the holding down studs. Gasket compound must not be used. Fit new 'O' rings to each cylinder spigot.

5   Position the piston rings so that their end gaps are out of line with each other and fit a piston ring clamp to each piston. It is highly recommended that ring clamps be used for cylinder block

13 Fault diagnosis

Symptom	Cause	Remedy
Engine gradually fades and stops	Fuel starvation	Check vent hole in filler cap. Sediment in filter bowl or float chamber. Dismantle and clean.
Engine runs badly. Black smoke from exhausts	Carburettor flooding	Dismantle and clean carburettor. Check for punctured float or sticking float needle.
Engine lacks response and overheats	Weak mixture Air cleaner disconnected or hose split Modified silencer has upset carburation	Check for partial block in carburettors. Reconnect or renew hose. Replace with original design.
Oil pressure warning light comes on	Lubrication system failure	Stop engine immediately. Trace and rectify fault before re-starting.
Engine gets noisy	Failure to change engine oil when recommended	Drain off old oil and refill with new oil of correct grade. Clean oil filter element.

Chapter 3 Ignition system

Refer to Chapter 7 for details of 1977 to 1983 models

Contents

Specifications

Sparking plugs

Make	NGK
Type	B—8ES (XS650C models BP—7ES)
Gap	0.6—0.7 mm (0.024 - 0.028 in)
Equivalents	Motorcraft AG 2 — XS 2 models
	AG 1 — XS650 models

Ignition timing

... ...   13°BTDC (15°BTDC on XS650C model)
(ignition fully retarded)

Advance range

... ...   10°BTDC at tickover
to 38°BTDC at full advance

Contact breaker gap

... ...   0.3 — 0.4 mm (0.012 — 0.016 in)

Ignition coil

Make	Hitachi
Output	10 Kv @ 4,000 RPM

1 General description

The Yamaha 650 twins are equipped with a conventional twin coil and contact breaker ignition system. The contact breaker assemblies are mounted in a separate housing on the left-hand side of the cylinder head, and are operated from a cam mounted on the end of the camshaft.

While the contact breaker points are closed, current flows to the primary windings of the coil concerned. As the contact breaker points separate, the low tension current is interrupted, inducing a high tension (HT) current in the secondary windings of the coil. The resulting HT current is fed to the relevant sparking plug, where it jumps across the electrode gap, igniting the fuel/air mixture.

2 Alternator: checking the output

1   The alternator is instrumental in creating the power in the ignition system, and any malfunction or failure will affect the operation of the ignition system. Should an alternator fault be indicated, reference should be made to Chapter 6, Section 2 for the output checking procedure.

3 Contact breakers: adjustment

1   To gain access to the contact breakers, the cover on the left-hand side of the camshaft should be removed. It is retained by two screws. The contact breakers are mounted on a plate in the housing, and are operated by a cam which protrudes through the plate.

2   Rotate the engine until one of the contact breaker fibre heels is at the highest point of the cam, and the points are at their widest opening. Examine the contact faces. If they are dirty, burnt or pitted, they should be removed for further attention as described in Section 4 of this Chapter. If the points are in sound condition, proceed with adjustment as follows:

3   If the gap between the contact breaker points is not within the permissible setting range of 0.3 - 0.4 mm (0.012 - 0.016 in) when measured with a feeler gauge, they should be adjusted. Slacken very slightly the cross point screw which locks the fixed contact adjustment. It should barely nip the contact support. Using a screwdriver, open or close the contact, tighten the lockscrew and check the setting. Repeat until correct.

4   When the satisfactory setting has been achieved, repeat the procedure for the other set of contacts. When both settings have been adjusted, spin the engine over a few times, and then recheck them.

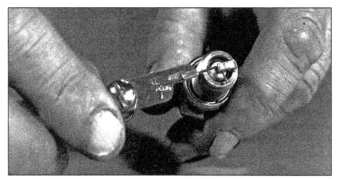

Electrode gap check - use a wire type gauge for best results

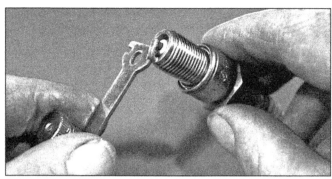

Electrode gap adjustment - bend the side electrode using the correct tool

Normal condition - A brown, tan or grey firing end indicates that the engine is in good condition and that the plug type is correct

Ash deposits - Light brown deposits encrusted on the electrodes and insulator, leading to misfire and hesitation. Caused by excessive amounts of oil in the combustion chamber or poor quality fuel/oil

Carbon fouling - Dry, black sooty deposits leading to misfire and weak spark. Caused by an over-rich fuel/air mixture, faulty choke operation or blocked air filter

Oil fouling - Wet oily deposits leading to misfire and weak spark. Caused by oil leakage past piston rings or valve guides (4-stroke engine), or excess lubricant (2-stroke engine)

Overheating - A blistered white insulator and glazed electrodes. Caused by ignition system fault, incorrect fuel, or cooling system fault

Worn plug - Worn electrodes will cause poor starting in damp or cold weather and will also waste fuel

4   Contact breaker points: removal, renovation and replacement

1   If the contact breaker points are found to be burnt, pitted or badly worn, they should be removed for dressing. If, however, it is necessary to remove a substantial amount of material before the faces can be restored, new contacts should be fitted.

2   The contact breaker assemblies are retained on a plate mounted in the left-hand housing. Before the retaining screw is removed, it is first necessary to detach the coil and condenser leads from the spring blade of the moving point. They are retained by a nut, which should be replaced to avoid displacing or losing the two insulating washers, after the spring blade has been removed from the support post.

3   Using a small electrical screwdriver, prise off the circlip which retains the moving contact assembly to its pivot pin. Remove the plain washer, followed by the moving contact complete with insulating washers. Make a note of the order in which components are removed, as they are easily assembled incorrectly.

4   The fixed contact is retained by a single screw located at the outer edge of the contact breaker base plate. **Note:** The left-hand cylinder contact breaker assembly (the lower of the two sets) is mounted on a half-plate, which in turn is attached to the main contact breaker base plate. On no account must the screws which retain these plates be disturbed, otherwise the ignition timing will be lost.

5   The points surfaces may be dressed by rubbing them on an oilstone or fine emery paper, keeping the points square to the abrasive surface. If possible, finish off by using Crokus paper to give a polished surface, which is less prone to subsequent pitting. Make sure all traces of abrasive are removed before reassembly.

6   Reassemble the contact breaker assembly by reversing the dismantling sequence, taking care that the insulating washers are replaced correctly. If this precaution is not observed, it is easy to inadvertently earth the assembly, rendering it inoperative. The pivot pin should be greased sparingly, and a few drops of oil applied to the cam lubricating wick.

7   If a contact breaker is being renewed due to excessive burning of the contacts, this is likely to have been caused by a faulty condenser. Refer to the next section if this is suspected.

5   Condenser: removal and replacement

1   There are two condensers contained in the ignition system, wired in parallel with the points. If a fault develops, ignition failure is likely to occur.

2   If the engine proves difficult to start, or misfiring occurs, it is possible a condenser is at fault. To check, separate the contact points by hand when the ignition is switched on. If a spark occurs across the points as they are separated by hand and they have a blackened and a burnt appearance, the condenser associated with that set of points can be regarded as unserviceable.

3   To test a condenser, sophisticated test equipment is necessary. In view of the small cost involved it is preferable to fit a new one, and observe the effect on engine performance by substitution.

4   The two condensers are clamped to the head steady plate and are wired individually to each set of points. The screws that hold the condensers to the plate also form the earth connection and should always be checked for tightness.

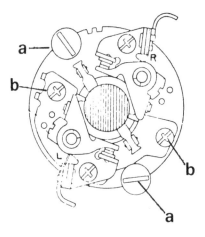

Fig. 3.1. Contact breaker gap and timing adjustment

1   *RH cylinder fixed contact lock screw*
2   *LH cylinder fixed contact lock screw*
3   *RH cylinder adjustment notch*
4   *LH cylinder adjustment notch*

A   *Contact breaker base plate locking scews*
B   *LH Contact breaker half plate locking screws*

3.3 Use screwdriver to move fixed contact setting

4.2 Do not slacken large outer screws (arrowed) or timing will be lost

6 Ignition coil - checking

1   The ignition coil is a sealed unit, designed to give long service without need for attention. It is located within the top frame tubes, immediately to the rear of the steering head assembly. If a weak spark and difficult starting causes the performance of the coil to be suspect, it should be tested by a Yamaha Agent or an auto-electrical expert who will have the appropriate test equipment. A faulty coil must be replaced; it is not possible to effect a satisfactory repair.

2   A defective condenser in the contact breaker circuit can give the illusion of a defective coil and for this reason it is advisable to investigate the condition of the condenser before condemning the the ignition coil. Refer to Section 5 of this Chapter for the appropriate details.

7 Ignition timing - checking and resetting

1   In order to check the accuracy of the ignition timing, it is first necessary to remove the contact breaker cover and the circular cover from the left-hand crankcase cover which gives access to the generator rotor. It will be observed that the stator is inscribed with two lines marked T and F and that there is a white timing mark on the rotor which also has an inscribed line.

2   If the ignition timing is correct, the F line will coincide exactly with the line scribed on the pointer, when the contact breaker points are just commencing to separate. In order to check both cylinders, it is necessary to turn the rotor one complete revolution (360°) so that the reading on the other cylinder is obtained. (See Chapter 1 Fig. 1.13).

3   If there is a minor discrepancy between the two readings, it is permissible to 'share' the error, so that both cylinders fire at exactly the same setting. Some variation is almost inevitable, unless the contact breaker cam has been manufactured to an un-usually high standard of accuracy. The magnitude of the error is however rarely sufficient to cause any problems, especially if the error is shared, as suggested. It should be noted that the F mark is, in fact, a small area contained by two scribed lines. These lines indicate the range through which the timing may safely vary between sets of contact breakers.

4   To adjust the position of the contact breaker points relative to the cam, the right-hand cylinder's points must set first. These are mounted directly on the circular baseplate. The left-hand points can be set next. These will be found to be mounted on a small half-plate, permitting independent adjust-ment. The contact breakers should be moved either clockwise or anti-clockwise until the points commence to separate as the timing marks on the rotor coincide.

5   It cannot be overstressed that optimum performance depends on the accuracy with which the ignition timing is set. Even a small error can cause a marked reduction in performance and the possibility of engine damage, as the result of overheating. The contact breaker gap must be checked and if necessary reset for both lobes of the cam BEFORE the accuracy of the ignition timing is verified since adjustments made at a later stage will affect the accuracy of the setting.

8 Automatic timing unit: examination and renovation

1   The automatic timing unit (ATU) is contained in a housing at the right-hand end of the camshaft, controlling the ignition timing mechanically by way of a rod which runs centrally in the camshaft. The assembly can be removed after the outer cover has been de-tached. Slacken the centre nut and remove the driving sleeve from the end of the rod, disengaging the operating weights from their slots. The ATU can be withdrawn after slackening the slotted nut which retains it by means of a small punch and hammer.

2   The unit comprises spring loaded balance weights, which move outwards against the spring tension as centrifugal force increases. The balance weights must move freely on their pivots and be

rust-free. The tension springs must also be in good condition. Keep the pivots lubricated and make sure the balance weights move easily, without binding. Most problems arise as a result of condensation within the engine, which causes the unit to rust and balance weight movement to be restricted.

3   When refitting the unit, the following points must be noted; The slotted driving sleeve is marked with an arrow denoting the direction of rotation. This must face outwards. The sleeve is also marked with a painted dot which should correspond with the paint mark on one of the bobweights.

9 Sparking plugs: checking and resetting the gaps

1   Two NGK B8ES or BP-7ES sparking plugs are fitted to the Yamaha 650 series as standard. Certain operating conditions may indicate a change in sparking plug grade, but generally the type recommended by the manufacturer gives the best all round service.

2   Check the gap of the plug points every three monthly or 2,000 mile service. To test the gap, bend the outer electrode to bring it closer to, or further away from the central electrode until a 0.028 in (0.7 mm) feeler gauge can be inserted. Never bend the centre electrode or the insulator will crack, causing engine damage if the particles fall into the cylinder whilst the engine is running.

3   With some experience, the condition of the sparking plug electrodes and insulator can be used as a reliable guide to engine operating conditions. See the accompanying diagram.

4   Always carry a spare pair of sparking plugs of the recommend-ed grade. In the rare event of plug failure, they will enable the engine to be restarted.

5   Beware of overtightening the sparking plugs, otherwise there is risk of stripping the threads from the aluminium alloy cylinder heads. The plugs should be sufficiently tight to seat firmly on their copper sealing washers, and no more. Use a spanner which is a good fit to prevent the spanner from slipping and breaking the insulator.

6   If the threads in the cylinder head strip as a result of over-tightening the sparking plugs, it is possible to reclaim the head by the use of a Helicoil thread insert. This is a cheap and convenient method of reclaiming the threads; most motorcycle dealers operate a service of this nature at an economic price.

7   Make sure the plug insulating caps are a good fit and have their rubber seals. They should also be kept clean to prevent tracking. These caps contain the suppressors that eliminate both radio and TV interference.

6.1 Coils are bolted either side of top tube

7.1 Points should separate at F mark. Left-hand line indicates full advance

7.3 LH timing is altered thus after slackening screws (arrowed)

10 Fault diagnosis

Symptom	Cause	Remedy
Engine will not start	Faulty ignition switch	Operate switch several times in case contacts are dirty. If lights and other electrics function, switch may need renewing.
	Starter motor not working	Discharged battery. Use kickstart until battery is recharged.
	Short circuit in wiring	Check whether fuse is intact. Eliminate fault before switching on again.
Engine misfires	Faulty condenser in ignition circuit	Renew condenser and retest.
	Fouled sparking plug	Renew plug and have original cleaned.
	Poor spark due to alternator failure and discharging battery	Check output from generator. Remove and recharge battery.
Engine lacks power and overheats	Retarded ignition timing	Check timing and also contact breaker gap. Check whether auto-advance mechanism has jammed.
Engine 'fades' when under load	Pre-ignition	Check grade of plugs fitted; use recommended grades only. Verify whether lubrication system has pressure.

Chapter 4 Frame and forks

Refer to Chapter 7 for details of 1977 to 1983 models

Contents

Specifications

Front forks

Type	Oil damped telescopic
Damping oil capacity	
XS1B	223 cc per leg
XS2, TX650	136 cc per leg
TX650A, XS650B, XS650C	155 cc per leg
Damping oil specification	SAE 10W/30 engine oil

Rear suspension

Type	Swinging arm

Rear suspension units

Type	Oil damped coil spring. Sealed oil content

---

1 General description

The Yamaha 650 series have a duplex tube frame of the full cradle type; that is, with the engine not comprising any part of the frame. Rear suspension is of the swinging arm type, using oil filled suspension units to provide the necessary damping action. The units are adjustable so that the spring ratings can be effectively changed within certain limits to match the load carried.

The front forks are of the conventional telescopic type, having internal, oil-filled dampers. The fork springs are contained within the fork stanchions and each fork leg can be detached from the machine as a complete unit, without dismantling the steering head assembly.

---

2 Front forks: removal from the frame

1 It is unlikely that the front forks will have to be removed from the frame as a complete unit, unless the steering head assembly requires attention or if the machine suffers frontal damage.
2 Commence front fork removal by disconnecting the main controls from the handlebars. In the case of machines fitted with a disc front brake, the hydraulic master cylinder must be removed. The master cylinder and operating lever is clamped to the bars by

two bolts. Disconnect the clutch cable and the decompressor cable (where fitted).
3 Remove the combined twistgrip/switch from the right-hand end of the handlebars, and the indicator/lighting switch from the the left-hand end of the bars.
4 Detach the headlamp glass and rim and disconnect the various snap connectors. The wires leading through to the handlebars can be pulled through the orifice in the rear of the headlamp, and the handlebar switches completely removed.
5 Slacken the four bolts that retain the handlebar clamps, and lift the handlebars away. Remove the tachometer and speedometer which are retained on a common bracket bolted to the top steering yoke. Disconnect the two drive cables by unscrewing the knurled ring on each cable end, and the indicator lamps.
6 Pull the two wires off the brake switch which is screwed into the hydraulic hose junction piece. The junction T piece is retained on the steering lower yoke by two bolts.
7 Remove the headlamp shell which is retained by two bolts, one of which passes through each fork shroud bracket into the side of the headlamp shell.
8 Place a sturdy support below the crankcase so that the front wheel is raised well clear of the ground. Remove the speedometer cable from the gearbox on the right-hand side of the wheel; tne cable is held by a screw ring.
9 Remove the split pin from the castellated wheel spindle nut, and

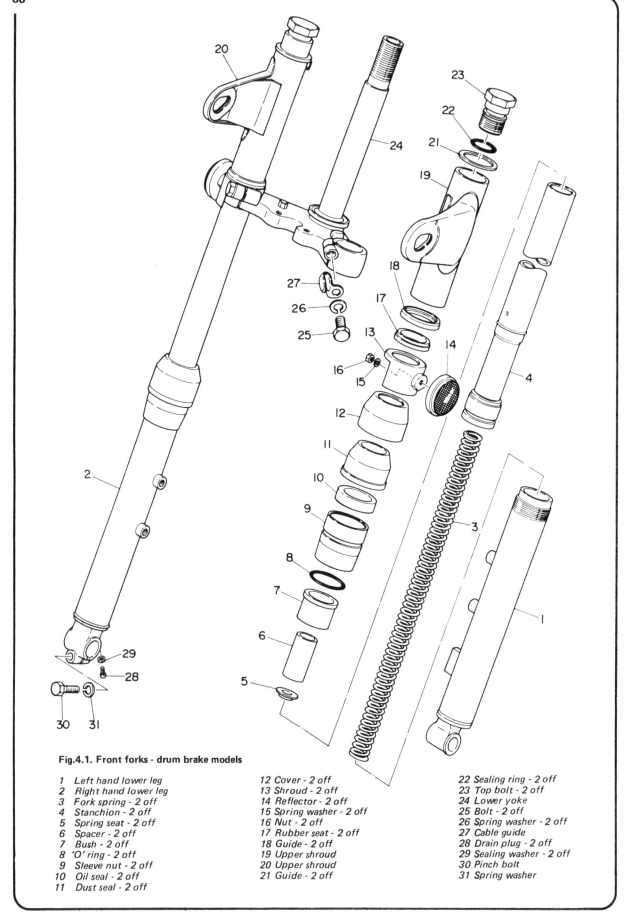

Fig.4.1. Front forks - drum brake models

1 Left hand lower leg	12 Cover - 2 off	22 Sealing ring - 2 off
2 Right hand lower leg	13 Shroud - 2 off	23 Top bolt - 2 off
3 Fork spring - 2 off	14 Reflector - 2 off	24 Lower yoke
4 Stanchion - 2 off	15 Spring washer - 2 off	25 Bolt - 2 off
5 Spring seat - 2 off	16 Nut - 2 off	26 Spring washer - 2 off
6 Spacer - 2 off	17 Rubber seat - 2 off	27 Cable guide
7 Bush - 2 off	18 Guide - 2 off	28 Drain plug - 2 off
8 'O' ring - 2 off	19 Upper shroud	29 Sealing washer - 2 off
9 Sleeve nut - 2 off	20 Upper shroud	30 Pinch bolt
10 Oil seal - 2 off	21 Guide - 2 off	31 Spring washer
11 Dust seal - 2 off		

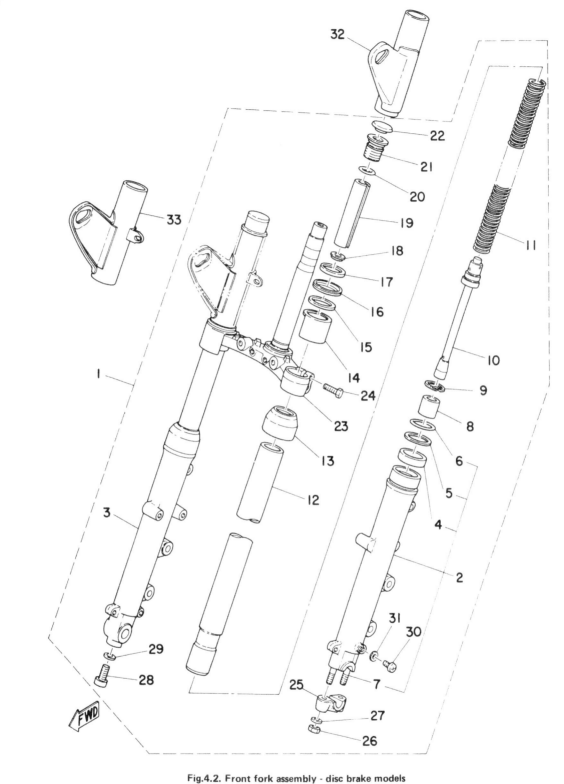

Fig.4.2. Front fork assembly - disc brake models

1 Front Fork assembly		18 Spring seat - 2 off	26 Nut - 2 off
2 Left hand lower fork leg	10 Damper assembly	19 Distance tube - 2 off	27 Plain washer - 2 off
3 Right hand lower fork leg	11 Spring - 2 off	20 'O'-ring - 2 off	28 Damper retaining screw - 2 off
4 Oil seal - 2 off	12 Stanchion - 2 off	21 Top bolt - 2 off	29 Sealing washer - 2 off
5 Washer - 2 off	13 Dust seal - 2 off	22 Blanking plug - 2 off	30 Drain screw - 2 off
6 Clip - 2 off	14 Shroud - 2 off	23 Lower yoke	31 Sealing washer - 2 off
7 Stud - 2 off	15 Sealing ring - 2 off	24 Pinch bolt - 2 off	32 Upper shroud (left hand)
8 Damper piston - 2 off	16 Spacer - 2 off	25 Spindle clamp	33 Upper shroud (right hand)
9 Circlip - 2 off	17 Guide - 2 off		

slacken off the nut. Loosen the two clamp retaining nuts on the bottom of the left-hand lower leg. Withdraw the wheel spindle using the tommy bar supplied in the tool kit, or a similar substitute. The wheel can now be disengaged from the brake caliper and removed.

Note: On machines fitted with twin front disc brakes, it will be found that it is impossible to pass the rim and tyre between the two calipers. In this case, remove the mudguard mounting bolts from one fork leg, to allow the caliper to be turned away to give clearance.

10 Remove the two bolts which retain the brake caliper assembly and remove the assembly together with the associate brake operating components. The brake hose is held by a clip and grommet and may be difficult to remove. It is better to bend the clip open to allow the brake hose and grommet to be freed than to try and force the hose out of the clip, which might cause damage.

11 When removing the various hydraulic brake parts it is important that care is taken not to bend or kink the brake hoses or pipes. Under no circumstance should the front brake be operated at any time during removal or there is danger of the caliper pistons being forced out of the cylinders with the resulting loss of fluid. Should this happen, the brake assembly must be bled after it has been replaced and the front wheel is in position. If any fluid is inadvertently spilled onto the paintwork it should be removed at once. Hydraulic brake fluid is a most effective paint remover.

12 Remove the front mudguard, which is retained by two bolts on the lower end of each fork leg, and by two bolts through brackets on the inside of each fork lower leg. Remove the mudguard complete with centre bracket and stays.

13 Unscrew the upper and lower pinch bolts which clamp the fork legs to the two fork yokes. Each fork leg can now be removed as a complete unit. It may be necessary to spring the yoke clamps apart with a screwdriver to allow the fork legs to be pulled down, out of position.

14 Loosen the pinch bolt which clamps the steering head stem and undo the crown domed bolt. Remove the thick washer and wave washer. With the aid of a soft-nosed mallet the fork top yoke can be tapped upwards and off the steering stem.

15 To release the lower yoke and the steering head stem, unscrew the adjuster ring at the top with a suitable 'C' spanner. If such a spanner is not available, a brass drift can be used to loosen the ring. As the steering head is lowered, the uncaged ball bearings from the lower race will be released, and care should be taken to catch them as they fall free. The bearings in the upper race will almost certainly stay in place.

16 It follows that much of this procedure can be avoided if it is necessary to remove the individual fork legs without disturbing the fork yokes and steering head bearings. Under these circumstances commence dismantling as described in paragraph 9 and work through to paragraph 13.

3 Front forks: dismantling

1 It is advisable to dismantle each fork leg separately using an identical procedure. There is less chance of unwittingly exchanging parts if this approach is adopted. Commence by draining the fork legs; there is a drain plug in each lower leg, located above the mudguard rear mounting stay.

2 Commence dismantling the fork leg by removing the Allen socket screw in the extreme lower end of the fork leg. Remove the chrome top bolt and pull out the spring. The fork leg may be held in the jaws of a vice in order to undo the chrome bolt, provided that the jaws are soft, or a length of inner tube is wrapped around the leg to protect it. It may be found in practice that the damper rod will turn in the lower leg when the socket screw is

turned. The damper rod head has flats milled on it, and a piece of tubing with the end flattened can be passed down through the fork stanchion to hold it. The overall diameter of the head is 12 mm, with the flats measuring 10 mm across. A piece of tubing having an internal diameter of ½ in proves ideal for this purpose (see photograph). **Note:** Early models were fitted with a slightly different type of fork, in which the fork top bush and seal are carried in a large diameter sleeve, effectively retaining the fork leg on the stanchion. In this case, slacken the sleeve using either a strap wrench, or if improvising, by using a worm drive hose clip tightened around the sleeve. The two halves can now be pulled apart. In the case of the later type of fork, finish dismantling as follows:

3 Prise the dust cover off the lower fork leg and pull the fork tube out of the lower leg. If the fork tube is now inverted the damper rod assembly will fall out. The various damper components can now be slipped off the damper rod for inspection.

4 Steering head bearings: examination and renovation

1 Before commencing reassembly of the forks, examine the steering head races. The ball bearing tracks of the respective cup and cone bearings should be polished and free from indentations, cracks or pitting. If signs of wear are evident, the cups and cones must be renewed. In order for the straight line steering on any motorcycle to be consistently good, the steering head bearings must be absolutely perfect. Even the smallest amount of wear on the cups and cones may cause steering wobble at high speeds and judder during heavy front wheel braking. The cups and cones are an interference fit on their respective seatings and can be tapped from position with a suitable drift.

2 Ball bearings are relatively cheap. If the originals are marked or discoloured they **must** be renewed. To hold the steel balls in place during reassembly of the fork yokes, pack the bearings with grease. Both the upper race and the lower race contain nineteen (19) ¼ inch ball bearings. Although space will be left for one extra steel ball, making the number up to twenty, an extra steel ball must not be fitted. The gap allows the bearings to work correctly, stopping them skidding and accelerating the rate of wear.

2.13 Slacken pinch bolts in yokes to release fork legs

2.14 Clamp retains the upper fork yoke

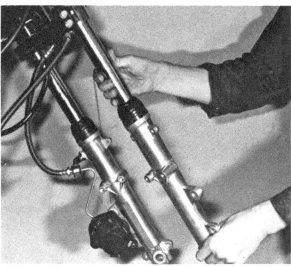

2.16 Fork legs can be removed individually if required

3.2a Note flats milled on damper rod end ...

3.2b ... can be held using flattened tubing ...

3.2c ... to allow screw to be slackened

3.3 Slide dust seal off, and withdraw stanchion

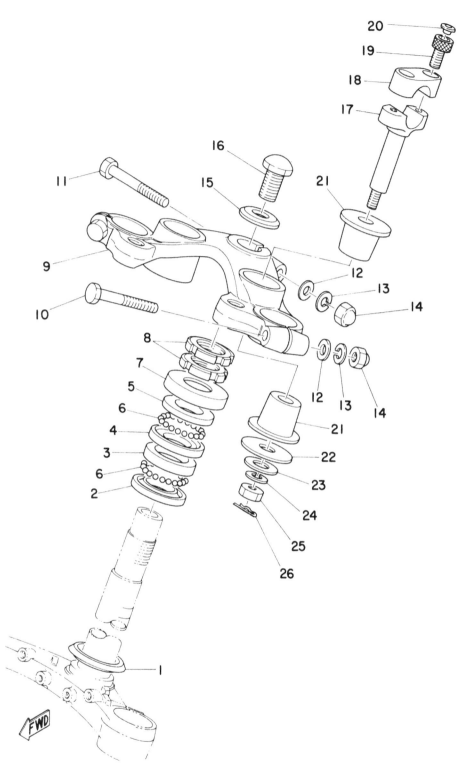

Fig. 4.3. Steering head assembly

1	Dust seal	8	Nut - 2 off	15	Washer
2	Lower bearing cup	9	Fork upper yoke	16	Bolt
3	Lower bearing cone	10	Pinch bolt - 2 off	17	Handlebar clamp -
4	Upper bearing cup	11	Bolt		lower - 2 off
5	Upper bearing cone	12	Washer - 3 off	18	Handlebar clamp - upper -
6	Ball bearing - 48 off	13	Spring washer - 3 off		2 off
7	Ball race cover	14	Domed nut - 3 off	19	Allen screw - 4 off

20	Cap bolt - 4 off
21	Rubber bushing - 4 off
22	Wasner - 2 off
23	Washer - 2 off
24	Spring washer - 2 off
25	Nut - 2 off
26	Clip - 2 off

5 Front forks: examination and renovation

1 The parts most liable to wear over an extended period of service are the internal surfaces of the lower leg and the outer surfaces of the fork stanchion. If there is excessive play between these two parts they must be replaced as a complete unit. Check the fork stanchion for scoring over the length which enters the oil seal. Bad scoring here will damage the oil seal and lead to fluid leakage.

2 It is advisable to renew the oil seals when the forks are dismantled even if they appear to be in good condition. This will save a strip-down of the forks at a later date if oil leakage occurs. The oil seal in the top of each lower fork leg is retained by an internal 'C' ring which can be prised out of position with a small screwdriver. Check that the dust excluder rubbers are not split or worn where they bear on the fork tube. A worn excluder will allow the ingress of dust and water which will damage the oil seal and eventually cause wear of the fork tube.

3 It is not generally possible to straighten forks which have been badly damaged in an accident, particularly when the correct jigs are not available. It is always best to err on the side of safety and fit new ones, especially since there is no easy means to detect whether the forks have been over stressed or metal fatigued. Fork stanchions can be checked, after removal from the lower legs, by rolling them on a dead flat surface. Any misalignment will be immediately obvious.

4 The fork springs will take a permanent set after considerable usage and will need renewal if the fork action becomes spongy. Compare the suspect springs with new components where possible. Always renew them as a matched pair.

5 Fork damping is governed by the viscosity of the oil in the fork legs, normally SAE 10/W 30, and by the action of the damper assembly. Each fork leg holds the following quantities of damping oil:

XSIB:	:	*223 cc*
XS2, TX650	:	*136 cc*
TX650A, XS650B, XS650C	:	*155 cc*

6 Front forks: replacement

1 Replace the front forks by following in reverse the dismantling procedures described in Section 2 and 3 of this Chapter. Before fully tightening the front wheel spindle clamps and the fork yoke pinch bolts, bounce the forks several times to ensure they work freely and are clamped in their original settings. Complete the final tightening from the wheel spindle clamps upward.

2 Do not forget to add the recommended quantity of fork damping oil to each leg before the bolts in the top of each fork leg are replaced. Check that the drain plugs have been re-inserted and tightened before the oil is added.

3 If the fork stanchions prove difficult to re-locate through the fork yokes, make sure their outer surfaces are clean and polished so that they will slide more easily. It is often advantageous to use a screwdriver blade to open up the clamps as the stanchions are pushed upward into position.

4 Before the machine is used on the road, check the adjustment of the steering head bearings. If they are too slack, judder will occur. There should be no detectable play in the head races when the handlebars are pulled and pushed, with the front brake applied hard.

5 Overtight head races are equally undesirable. It is possible unwittingly to apply a loading of several tons on the head bearings by overtightening, even though the handlebars appear to turn quite freely. Overtight bearings will cause the machine to roll at low speeds and give generally imprecise handling with a tendency to weave. Adjustment is correct if there is no perceptible play in the bearings and the handlebars will swing to full lock in either direction, when the machine is on the centre stand with the front wheel clear of the ground. Only a slight tap should cause the handlebars to swing.

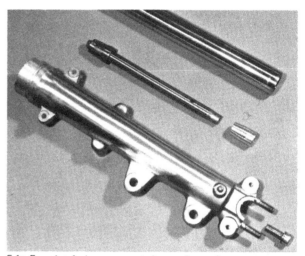

5.1a Examine fork components for scoring and wear

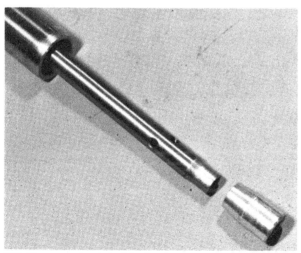

5.1b Damper rod seats in this alloy boss

5.1c Wear in these components will diminish damping effect

5.2 Lower leg seal is retained by wire circlip

5.4 Renew springs if these have taken permanent set

6.1a Lubricate seal lip before fitting stanchion

6.1b Refit front wheel, and bounce forks a few times ...

6.1c ... before tightening pinch bolts finally

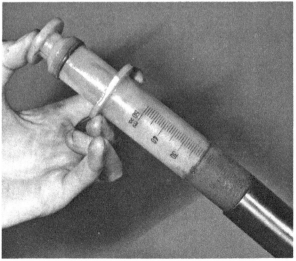

6.2a Refil each fork leg with correct quantity of oil ...

6.2b ... but make sure drain plug is refitted first

6.2c Check the condition of top bolt 'O' ring ...

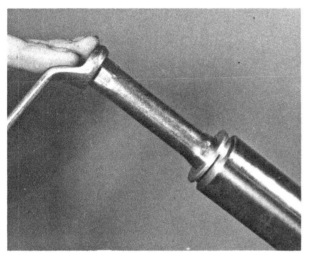

6.2d ... before tightening it as shown here

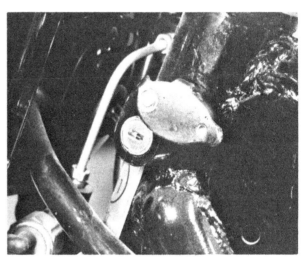

7.1 Steering lock is located on LH side of headstock

7 Steering head lock

1   The steering lock mechanism is mounted in a welded body attached to the steering head. When the lock is operated, a pin passes into the steering head and engages with a slot in the steering column, effectively locking the steering.

2   With the exception of occasional lubrication with light machine oil, little can be done in the way of maintenance. Should the lock malfunction, it will be necessary to renew it, and to obtain new keys to suit.

8 Frame: examination and renovation

1   The frame is unlikely to require attention unless accident damage has occurred. In some cases, replacement of the frame is the only satisfactory course of action if it is badly out of alignment. Only a few frame repair specialists have the jigs and mandrels necessary for resetting the frame to the required standard of accuracy and even then there is no easy means of assessing to what extent the frame may have been overstressed.

2   After the machine has covered a considerable mileage, it is advisable to examine the frame closely for signs of cracking or

splitting at the welded joints. Rust can also cause weakness at these joints. Minor damage can be repaired by welding or brazing, depending on the extent and nature of the damage.

3   Remember that a frame which is out of alignment will cause handling problems and may even promote speed wobbles. If misalignment is suspected, as the result of an accident, it will be necessary to strip the machine completely so that the frame can be checked and, if necessary, renewed.

9 Swinging arm rear fork: dismantling, examination and renovation

1   The rear fork assembly pivots on a detachable bush within each end of the fork crossmember and a pivot shaft which itself is surrounded by two bushes separated by a long distance piece. The pivot shaft passes through frame lugs on each side of the engine unit, and the two centre bushes and distance piece, so that the inner and outer bushes form the bearing surfaces. It is quite easy to renovate the swinging arm when wear necessitates attention.

2   To remove the rear swinging arm fork, first position the machine on the centre stand so that it rests firmly and securely Remove the final drive chain by detaching the master link, and

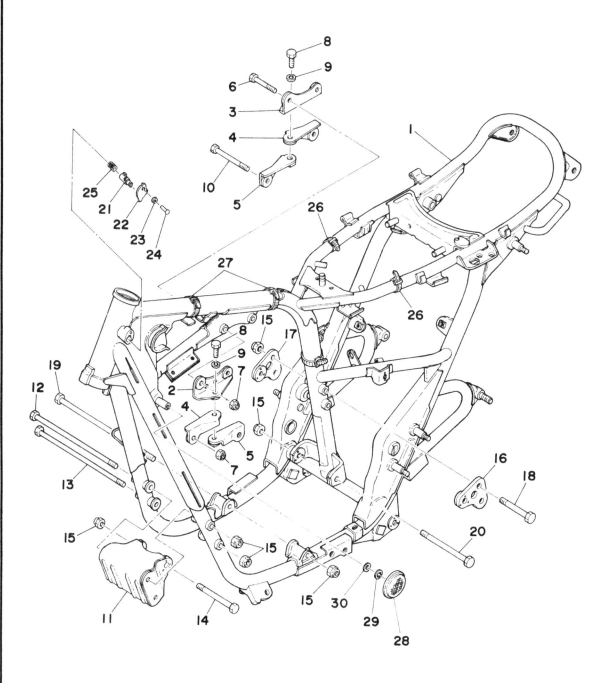

Fig. 4.4. Frame assembly

1 Frame assembly complete	6 Bolt - 2 off	15 Self locking nut - 8 off	23 Wave washer
2 Left hand upper head steady plate	7 Self locking nut - 4 off	16 Engine rear mounting plate	24 Rivet
3 Right hand upper head steady plate	8 Bolt - 2 off	17 Engine rear mounting plate	25 Spring
4 Lower head steady plate halves - 2 off	9 Spring washer - 2 off	18 Bolt - 3 off	26 Wiring strap - 2 off
5 Lower head steady plate halves - 2 off	10 Bolt - 2 off	19 Bolt	27 Wiring strap - 2 off
	11 Engine front mounting plate	20 Bolt	28 Reflector - 2 off
	12 Bolt	21 Steering lock assembly	29 Spring washer - 2 off
	13 Bolt	22 Lock cap	30 Plain washer - 2 off
	14 Bolt		

then unscrew the two crosshead screws that hold the chainguard.

3   Detach the brake torque arm from the lug on the brake plate and from the lug on the frame. In both cases the torque arm bolt retaining nuts are secured by split pins. Unscrew the adjuster ring on the brake operating arm and pull the rod through the trunnion on the brake arm. Replace the adjuster ring to avoid the loss of the brake rod spring.

4   Remove the split pin from the end of the wheel spindle and remove the castellated nut. The wheel spindle can now be pulled out to the left. Knock the wheel spacer out of position from between the brake plate and the fork end. The rear wheel can now be pulled clear of the swinging arm fork. Note that late XS650C models have a metal fillet retained by a bolt at each fork end. These can be removed after the adjusters have been swung clear.

5   Detach both rear suspension units at their lugs on the swinging arm fork. Each unit is held by a single bolt screwed into a threaded hole in the swinging arm.

6   Remove the locknut from the end of the pivot shaft, which can then be tapped out from the left-hand side. Working the swinging arm fork up and down will aid removal of the shaft. The swinging arm fork is now free to be pulled from position between the two frame lugs.

7   Remove the dust excluder caps from each end of the fork

crossmember. Note the presence of the two shims in the caps. Push out the outer shaft.

8   Wash the crossmember and shaft in petrol or another solvent. Do not remove the outer bushes from position in the fork crossmember unless they need renewal as they are made of a brittle material that will probably fracture while being drifted out. With a micrometer or vernier gauge, measure the internal diameter of the bushes and the outside diameter of the shaft. If any component is outside the service limit, the bearings should be renewed as a complete set. Check the pivot for straightness by rolling it on the edge of a dead flat surface. If the shaft is bent it must be renewed or straightened.

9   Reassemble the swinging arm fork by reversing the dismantling procedure. Grease the pivot shaft and bearings liberally before reassembly and check that the shims in the dust caps are in good condition.

10 Worn swinging arm pivot bearings will give imprecise handling with a tendency for the rear end of the machine to twitch or hop. The play can be detected by placing the machine on its centre stand and with the rear wheel clear of the ground, pulling and pushing on the fork ends in a horizontal direction. Any play will be magnified by the leverage effect. In the UK, excess play will cause the machine to fail its DoT test.

9.6 Remove lower rear suspension mounting bolts

9.7a It is advisable to detach the chainguard

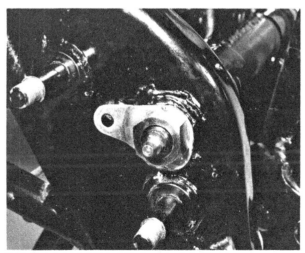

9.7b Knock back tab and slacken pivot shaft nut

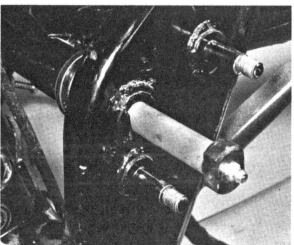

9.7c Shaft can be withdrawn from left hand side ...

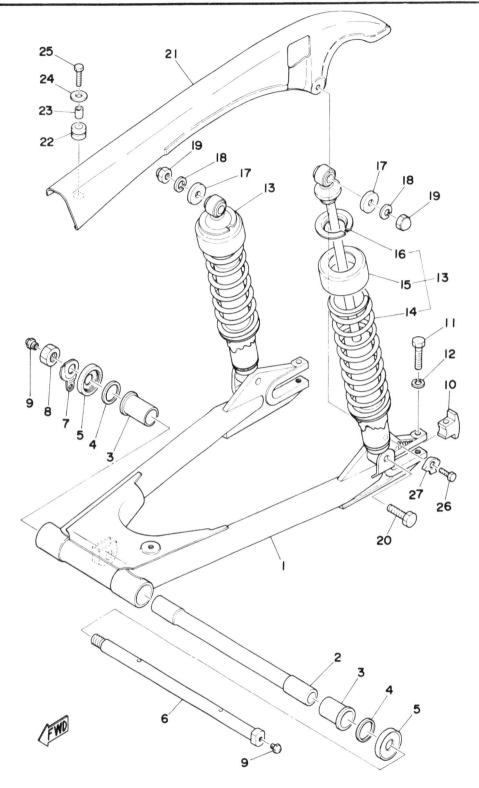

Fig. 4.5. Swinging arm rear suspension

1 Swinging arm fork	7 Lock washer	14 Rear suspension unit spring - 2 off
2 Swinging arm bush - centre	8 Nut	15 Upper shroud - 2 off
3 Swinging arm bush - end - 2 off	9 Grease nipple - 2 off	16 Upper spring seat - 2 off
4 Shim - thickness as required	10 Fillet - 2 off	17 Washer - 2 off
5 Dust cover - 2 off	11 Bolt - 2 off	18 Spring washer - 2 off
6 Pivot shaft	12 Spring washer - 2 off	19 Domed nut - 2 off
	13 Rear suspension unit - 2 off	20 Bolt - 2 off
		21 Chainguard
		22 Grommet
		23 Collar
		24 Washer
		25 Bolt
		26 Bolt
		27 Washer

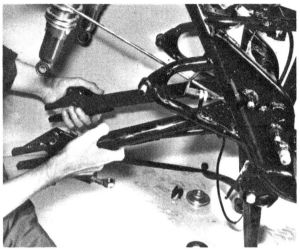

9.7d ... allowing swinging arm to be removed

9.8a Ends caps contain shims - do not lose them

9.8b Outer shaft can be displaced from crossmember

9.9 Do not remove bushes unless renewal is necessary

10 Rear suspension units: examination

1 The rear suspension units fitted to the Yamaha 650 machine are of the normal hydraulically damped type, adjustable to give five different spring settings. A 'C' spanner included in the tool kit should be used to turn the lower spring seat and so alter its position on the adjustment projection. When the spring seat is turned so that the effective length of the spring is shortened the suspension will become heavier.
2 If a suspension unit leaks, or if the damping efficiency is reduced in any other way the two units must be replaced as a pair. For precise roadholding it is imperative that both units re-act to movement in the way. It follows that the units must always be set at the same spring loading.

11 Centre stand: examination

1 The centre stand pivots on two shouldered bolts which pass through lugs on the frame. The pivot assemblies on centre stands are often neglected with regard to lubrication and this will eventually lead to wear. It is prudent to remove the pivot bolts from time to time and grease them thoroughly. This will prolong the effective life of the stand.
2 Check that the return spring is in good condition. A broken or weak spring may cause the stand to fall whilst the machine is being ridden, and catch in some obstacle, unseating the rider.

12 Prop stand: examination

1 The prop stand is attached to a lug welded to the left-hand lower frame tube. An extension spring anchored to the frame ensures that the stand is retracted when the weight of the machine is taken off the stand.
2 As with the centre stand, it is important to check that the stand and its mounting bolt are in sound condition, and that it is kept lubricated. After extended periods of use, the pivot bolt may wear, due to the leverage imposed upon it. If this occurs, it must be renewed before it wears the mounting plate through which it passes.

13 Footrests: examination

1   The footrests comprise two assemblies attached to the frame and can be detached as complete units. They are rubber-mounted to damp out engine vibration. The assembly is retained on rubber-sleeved studs by two nuts on each side.
2   Damage is likely only in the event of the machine being dropped. Slight bending can be rectified by stripping the assembly to the bare footrest bar, and straightening the bends by clamping the bar in a vice. A blowlamp should be applied to the affected area to avoid setting up stresses in the material, which may lead to subsequent fracturing.

14 Rear brake pedal: examination and renovation

1   The rear brake pedal is mounted on a spindle, which is supported in a bore in the frame lug. The pedal is held on its spline by a clamp bolt.
2   Should the pedal become bent in an accident, it can be straightened in a similar manner to that given for footrests in the preceding Section. If severely distorted, it should be renewed.

15 Dualseat: removal and replacement

1   The dualseat is attached to the right-hand side of the frame by two pivot pins, on which it hinges. A key operated latch on the left-hand side locks the seat in position, and also releases the helmet lock.
2   To remove the seat from the machine, unlock the seat latch and hinge the seat upwards. Remove the two R-shaped spring pins and push out the two small clevis pins on which the seat pivots. The hinge is bolted to the steel seat base, the other half being welded to the frame.

16 Kickstart lever: examination and renovation

1   The kickstart lever is splined and is secured to its shaft by means of a pinch bolt. The kickstart crank swivels so that it can be tucked out of the way when the engine is started. It is held in position on the swivel by a washer and circlip. A spring-loaded ball bearing locates the kickstart arm in either the operating or folded position; if the action becomes sloppy it is probable that the spring behind the ball bearing needs renewing. It is advisable to remove the circlip and washer occasionally, so that the kickstart crank can be detached and the swivel greased.
2   It is unlikely that the kickstart crank will bend in an accident unless the machine is ridden with the kickstart in the operating

and not folded position. It should be removed and straightened, using the same technique as that recommended for the footrests in Section 13.2.

17 Speedometer and tachometer heads: removal and replacement

1   The speedometer and tachometer heads are rubber-mounted and attached to the top fork yoke by means of two chromium-plated domed nuts at the base of each instrument. If the nuts are removed and the drive cable detached, the head can be lifted away. It will be necessary to remove the bulbs from the base of each instrument head by pulling the bulb-holders from their seatings; each is retained by a rubber cup.
2   The instruments are retained to each instrument case by anti-vibration sleeves at the top of the cases. Do not misplace these sleeves which are interposed between the instrument and the instrument case to damp out the undesirable effects of vibration.
3   Apart from defects in either the drive or the drive cable, a speedometer or tachometer that malfunctions is difficult to repair. Fit a new one, or alternatively entrust the repair to a competent instrument repair specialist.
4   Remember that a speedometer in correct working order is a statutory requirement in the UK. Apart from this legal requirement, reference to the odometer reading is the best means of keeping in pace with the maintenance schedule.

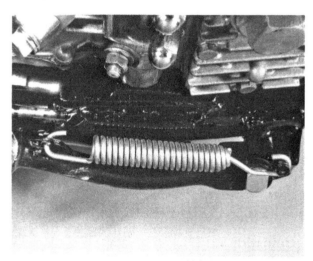

12.1 Examine prop stand for wear, and lubricate

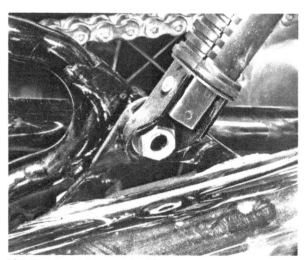

13.1 Rear footrests are retained by nut

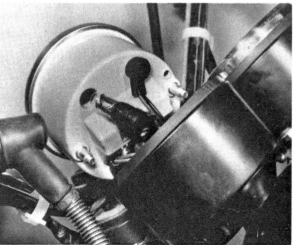

17.1 Withdraw bulbholders as instrument is removed

18 Speedometer and tachometer drive cables: examination and maintenance

1   It is advisable to detach both cables from time to time in order to check whether they are lubricated adequately, and whether the outer coverings are compressed or damaged at any point along their run. Jerky or sluggish movements can often be attributed to a cable fault.

2   For greasing, withdraw the inner cable. After wiping off the old grease, clean with a petrol-soaked rag and examine the cable for broken strands or other damage.

3   Regrease the cable with high melting point grease, taking care not to grease the last six inches at the point where the cable enters the instrument head. If this precaution is not observed, grease will work into the head and immobilise the movement.

4   If either instrument ceases to function, suspect a broken cable. Inspection will show whether the inner cable has broken; if so, the the inner cable alone can be renewed and reinserted in the outer casing, after greasing. Never fit a new inner cable alone if the outer covering is damaged or compressed at any point.

19 Speedometer and tachometer drives: location and examination

1   In the case of the disc front brake machines, the speedometer drive gearbox is fitted on the left-hand side of the front wheel hub. On drum front brake machines the gearbox is an integral part of the brake plate and is driven internally from the front hub. In both cases the drive rarely gives trouble provided it is kept properly lubricated. Lubrication should take place whenever the front wheel is removed for wheel bearing inspection or replacement.

2   The tachometer drive is taken from the right-hand outer casing, which houses the drive shaft. Drive is taken from a crankshaft mounted pinion which also provide drive to the oil pump. The drive assembly can be removed after slackening the single external retaining bolt. The unit should be greased sparingly during reassembly, but warrants no other attention.

20 Cleaning the machine

1   After removing all surface dirt with a rag or sponge which is washed frequently in clean water, the machine should be allowed to dry thoroughly. Application of car polish or wax to the cycle parts will give a good finish, particularly if the machine receives this attention at regular intervals.

2   The plated parts should require only a wipe with a damp rag, but if they are badly corroded, as may occur during the winter when the roads are salted, it is permissible to use one of the proprietary chrome cleaners. These often have an oily base which will help to prevent corrosion from recurring.

3   If the engine parts are particularly oily, use a cleaning compound such as Gunk or Jizer. Apply the compound whilst the parts are dry and work it in with a brush so that it has an opportunity to penetrate and soak into the film of oil and grease. Finish off by washing down liberally, taking care that water does not enter the carburettors, air cleaners or the electrics. If desired, the now clean aluminium alloy parts can be enhanced still further when they are dry by using a special polish such as Solvol Autosol. This will restore the full lustre.

4   If possible, the machine should be wiped down immediately after it has been used in the wet, so that it is not garaged under damp conditions which will promote rusting. Make sure that the chain is wiped and re-oiled, to prevent water from entering the rollers and causing harshness with an accompanying rapid rate of wear. Remember there is less chance of water entering the control cables and causing stiffness if they are lubricated regularly as described in the Routine Maintenance Section.

21 Fault diagnosis - frame and forks

Symptom	Cause	Remedy
Machine veers either to the left or the right with hands off handlebars	Bent frame Twisted forks Wheels out of alignment	Check and renew. Check and renew. Check and realign.
Machine rolls at low speed	Overtight steering head bearings	Slacken until adjustment is correct.
Machine judders when front brake is applied	Slack steering head bearings Worn forks	Tighten until adjustment is correct. Dismantle forks and renew worn parts.
Machine pitches on uneven surfaces	Ineffective fork dampers Ineffective rear suspension units Suspension too soft	Check oil content. Check whether units still have damping action. Raise suspension unit adjustment one notch.
Fork action stiff	Fork legs out of alignment (twisted in yokes)	Slacken yoke clamps, and fork top bolts. Pump fork several times then retighten from bottom upwards.
Machine wanders. Steering imprecise Rear wheel tends to hop	Worn swinging arm pivot	Dismantle and renew bushes and pivot shaft.

Chapter 5 Wheels, brakes and tyres

Refer to Chapter 7 for details of 1977 to 1983 models

Contents

Specifications

Tyres

Front	3.50 in — 19 in — 4PR
Rear	4.00 in — 18 in — 4PR

Tyre pressures

Front	23—25 psi (1.6 kg/cm^2)
Rear	28—30 psi (2.0 kg/cm^2)

Brakes

Front	(XS1B): 192 mm twin leading shoe drum brake
	(other models): single or twin hydraulic disc brake
Rear	180 mm single leading shoe drum brake

1 General description

All of the Yamaha 650 twins are equipped with a 19 in steel-rimmed front wheel carrying a 3.50 in. section tyre. The 18 in rear wheel has a 4.00 in. section tyre.

The XS1B models are fitted with a drum front brake of twin leading show design (TLS). Later models are equipped with a single or double hydraulic disc brake.

The rear brake on all models is a single leading shoe (SLS) drum brake.

2 Wheels: examination and renovation

1 Place the machine on its centre stand and chock the crankcase underside so that the wheel to be examined is raised clear of the ground. Spin the wheel and check the rim alignment, noting that in the case of the rear wheel it is preferable to disconnect the drive chain so that the wheel can spin freely. Small irregularities can be corrected by tightening the spokes in the affected area although a certain amount of experience is necessary to prevent over-correction. Any flats in the wheel rim will be evident at the same time. These are more difficult to remove and in most cases, it will be necessary to have the wheel rebuilt on a new rim. Apart from the effect on stability, a flat will expose the tyre bead and walls to greater risk of

damage if the machine is run with a deformed wheel.

2 Check for loose and broken spokes. Tapping the spokes is the best guide to tension. A loose spoke will produce a quite different sound and should be tightened by turning the nipple in an anti-clockwise direction. Always check for run-out by spinning the wheel again. If the spokes have to be tightened by an excessive amount, it is advisable to remove the tyre and tube as detailed in Section 21 of this Chapter. This will enable the protruding ends of the spokes to be ground off, thus preventing them from chafing the inner tube and causing punctures.

3 Front wheel disc brake: examination and renovation

1 Check the front brake master cylinder, hose and caliper unit for signs of fluid leakage. Pay particular attention to the condition of the synthetic rubber hose, which should be renewed without question if there are signs of cracking, splitting or other exterior damage.

2 Check the level of hydraulic fluid by removing the cap on the brake fluid reservoir and lifting out the diaphragm and diaphragm plate. This is one of the maintenance tasks, which should never be neglected. Make certain that the handlebars are in the central position when removing the reservoir cap, because if the fluid level is high., the fluid will spill over the reservoir brim. If the level is particularly low, the fluid delivery passage

will be allowed direct contact with the air and may necessitate the bleeding of the system at a later date. A level mark is given on the inside of the reservoir cylinder; if the level is below the mark, brake fluid of the correct grade must be added. **NEVER USE ENGINE OIL** or anything other than the recommended fluid. Other fluids have unsatisfactory characteristics and will quickly destroy the seals.

3 The brake pads can be checked for wear in position, by measuring the gap between the disc surface and the small tang projecting from the pad backing plate with the front brake lever squeezed. This gap must measure at least 0.5 mm (0.02 in) when checked with feeler gauges. The nominal overall pad thickness, including the backing plate, is 9mm (0.354 in), the overall wear limit being 4.5 mm (0.177 in). On no account must the pads be allowed to go below the specified limit. There is a very real danger of the disc being scored, and consequently needing renewal, if this happens.

4 The pads can be detached for renewal, either by removing the front wheel as described in Section 11 or by removing the caliper mounting bolts and tilting the assembly upwards to clear the disc. It is recommended that the latter course of action be adopted. If the caliper is detached from the fork leg, it should be supported so as not to place any strain on the hydraulic hose.

5 Use a small screwdriver to displace the old pads, taking care not to damage the caliper unit. It may be found that there is insufficient clearance for the new pads during reassembly, in which case it will be necessary to push the pistons inwards. This can be accomplished after slackening the bleed nipple slightly, to allow the hydraulic fluid to be displaced..**Important:** Note that whenever the hydraulic system is disturbed in any way, it is **essential** that the system is bled after reassembly, to obviate any risk of air being trapped in the system. See Section 8 of this Chapter. Care should also be taken to ensure that the hydraulic fluid is not allowed to come into contact with any plastic or painted parts of the machine.

6 Ensure that the new pads seat correctly in the caliper. When refitting the caliper to the fork leg, take care not to damage the pads of the edge of the disc. Tighten the retaining bolts to 4.0 - 5.0 m - kg (29 - 36 ft - lbs).

4 Front wheel disc brake: caliper unit - examination and renovation

1 The caliper or calipers employed on the Yamaha 650 twins

3.4 Remove caliper to gain access to pads

are of the double piston type and are bolted rigidly to the fork leg(s).Any trace of fluid leakage, or sponginess in the brakes s warrants immediate investigation. As a general rule, the caliper seals fail more often than those of the master cylinder, so the caliper(s) should be looked at first.

2 Start by slackening the gland nut which retains the brake pipe to the caliper unit, and freeing the pipe. To prevent fluid loss from the reservoir, pull the handlebar lever in and secure it with an elastic band. It is wise to cover the end of the pipe to prevent the ingress of contaminants.

3 Slacken and remove the caliper unit mounting bolts, and lift the unit away. Remove the pads as described in the preceding Section. Before any further dismantling is undertaken, prepare a clean area in which to work. It is most important that no foreign matter is allowed to get inside the caliper, therefore thorough cleaning of the outside of the unit is essential.

4 Remove the two bolts which clamp the caliper halves together, and separate the two halves. Wrap each caliper half in a clean rag, and displace the piston, using a compressed air jet. Prise out the seals from each caliper half.

5 Examine the piston for signs of scoring. This is normally caused by abrasive contaminants getting into the hydraulic fluid and becoming trapped in the seal surfaces. If any scoring is apparent, the piston must be renewed along with the seals. It is a wise precaution to renew the seals each time the caliper is dismantled. Their relatively low cost does not justify re-use.

6 Reassemble the caliper halves by reversing the dismantling sequence. Before installing the pistons, lubricate the seals thoroughly with clean hydraulic fluid. Fit a **new** caliper sealing ring in the fluid port between the two halves. The manufacturers recommend that new bridge bolts are used to retain the two halves. Tighten these to 0.6 - 1.0 m - kg (4.24 - 7.23 ft - lbs). The system must be bled of air after completion. See Section 8.

5 Front wheel disc brake: disc(s) - examination and renovation

1 The brake disc, bolted to the right-hand side of the front wheel hub, rarely requires attention. Check for run out, which may have occurred as the result of crash damage, and for wear, Run out should not exceed 0.15 mm (0.006 in) at any point, and the disc itself must not be permitted to wear below the limit thickness of 6.5 mm. If these figures are exceeded in either case, the disc must be renewed.

2 To remove the disc from the wheel, it is first necessary to detach the front wheel from the machine. (See Section 11). A small wooden wedge can be inserted in the caliper between the brake pads, to prevent them being expelled if the brake lever is operated inadvertently. There is in any case a tendency to creep in hydraulic systems, producing the same results.

3 The disc itself is attached to a cast boss. The eight mounting bolts are locked in position by spring washers, which must always be replaced on reassembly. The disc surfaces may now be examined for scoring, which is usually caused by particles of grit becoming embedded in the friction material of the pads. If excessive, the disc should be renewed as deep scoring is detrimental to braking efficiency. Reassembly is a direct reversal of the dismantling procedure, ensuring that each component is thoroughly cleaned, especially the mating surfaces, to obviate the risk of misalignment.

6 Master cylinder - examination and renovation

1 The master cylinder is unlikely to give trouble unless the machine has been stored for a lengthy period or until a considerable mileage has been covered. The usual signs of trouble are leakage of hydraulic fluid and a gradual fall in the fluid reservoir content.

2 To gain full access to the master cylinder, commence the dismantling operation by attaching a bleed tube to the caliper unit bleed nipple. Open the bleed nipple one complete turn, then

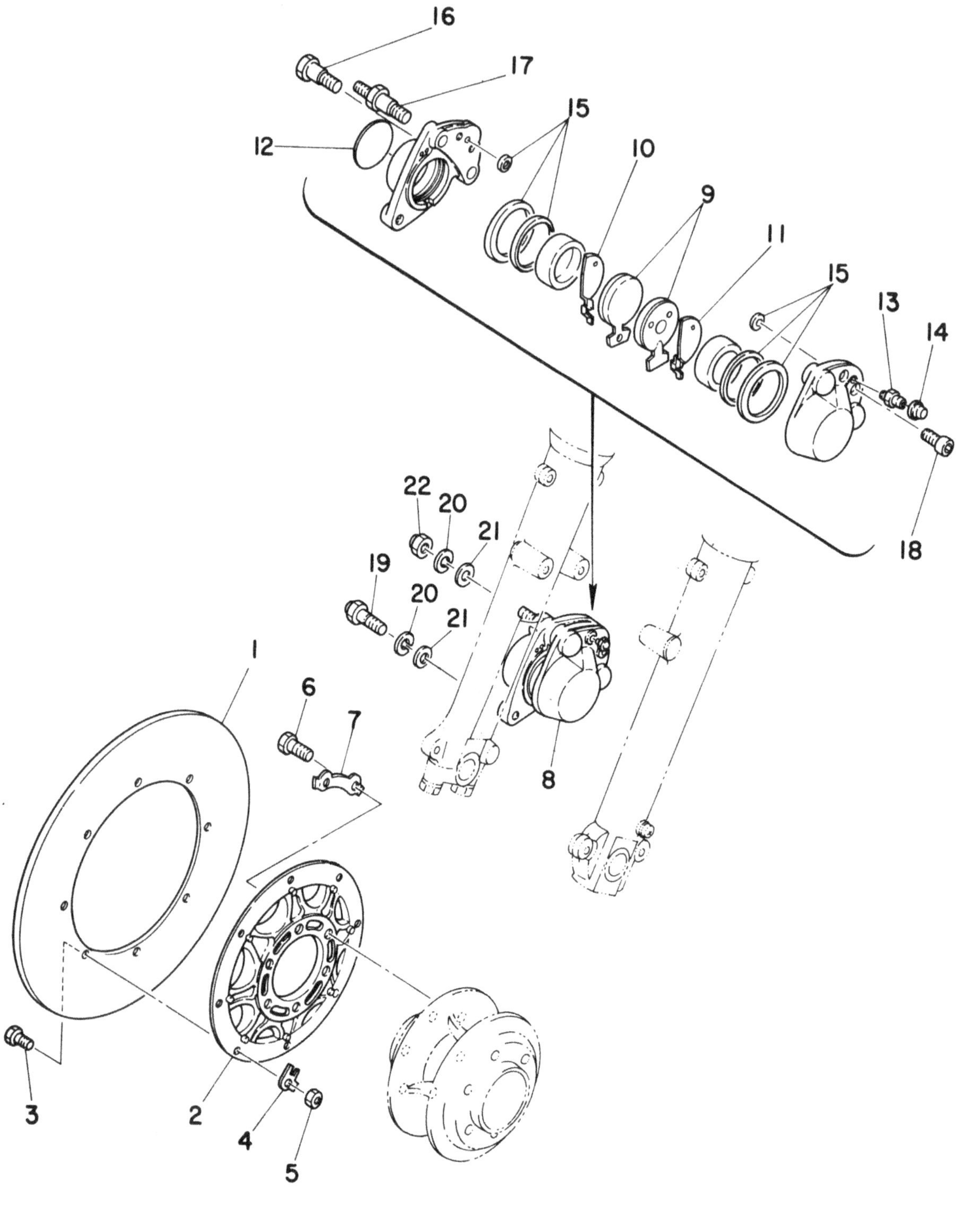

Fig.5.1. Front wheel disc brake caliper

1 Brake disc	9 Brake pad - 2 off	16 Bridge bolt 1
2 Disc mounting bracket	10 Shim 1	17 Bridge bolt 2
3 Bolt - 8 off	11 Shim 2	18 Bridge bolt 3 - 2 off
4 Lock washer - 8 off	12 Name plate	19 Bolt
5 Nut - 8 off	13 Bleed screw	20 Spring washer - 2 off
6 Bolt - 6 off	14 Cap	21 Plain washer - 2 off
7 Lock washer - 3 off	15 Caliper seal kit	22 Domed nut
8 Caliper assembly		

operate the front brake lever until all the fluid is pumped out of
the reservoir. Close the bleed nipple, detach the tube and store
the fluid in a closed container for subsequent re-use.

3   Detach the hose and also the stop lamp switch. Remove the
handlebar lever pivot bolt and the lever itself.

4   Access is now available to the piston and the cylinder and it
is possible to remove the piston assembly, together with all the
relevant seals. Take note of the way in which the seals are
arranged because they must be replaced in the same order.
Failure to observe this necessity will result in brake failure.

5   Clean the master cylinder and piston with either hydraulic
fluid or alcohol. On no account use either abrasives or other
solvents such as petrol. If any signs of wear or damage are
evident, renewal is necessary. It is not practicable to reclaim
either the piston or the cylinder bore.

6   Soak the new seals in hydraulic fluid for about 15 minutes
prior to fitting, then reassemble the parts **IN EXACTLY THE
SAME ORDER**, using the reversal of the dismantling procedure.
Lubricate with hydraulic fluid and make sure the feather edges
of the various seals are not damaged.

7   Refit the assembled master cylinder unit to the handlebar,
and reconnect the handlebar lever, hose, stop lamp switch etc.
Refill the reservoir with hydraulic fluid and bleed the entire
system by following the procedure detailed in Section 8 of this
Chapter.

8   Check that the brake is working correctly before taking the
machine on the road, to restore pressure and align the pads
correctly. Use the brake gently for the first 50 miles or so to
enable all the new parts to bed down correctly.

9   It should be emphasised that repairs to the master cylinder
are best entrusted to a Yamaha Agent, or alternatively, that the
defective part should be replaced by a new unit. Dismantling and
reassembly requires a certain amount of skill and it is imperative
that the entire operation is carried out under cleaner than
average conditions.

7   Front disc brake: hydraulic hose examination

1   An external hose is employed to transmit the hydraulic
pressure from the master cylinder to the caliper unit when
pressure is applied to the front brake lever.

2   The hose, of the flexible armoured type, must withstand
considerable pressure in service, and whilst it is easily ignored it
should be checked carefully as a sudden failure can be potentially
fatal. Look not only for signs of chafing against the wheel or
fork leg, but also for any stains due to fluid seeping from cracks
in the hose or from the connections at either end.

8   Front disc brake: bleeding the hydraulic system

1   Removal of the air from the hydraulic system is essential
to the efficiency of the braking system. Air can enter the
system due to leaks or when any part of the system has been
dismantled for repair or overhaul. Topping the system up will
not suffice, as air pockets will still remain, even small amounts
causing dramatic loss of brake pressure.

2   Check the level in the reservoir, and fill almost to the top.
Again, beware of spilling the fluid on to painted or plastic
surfaces.

3   Place a clean jar below the brake caliper unit and attach a
clear plastic tube from the caliper bleed screw to the container
Place some clean hydraulic fluid in the container so that the
pipe is always immersed below the surface of the fluid.

4   Unscrew the bleed screw one complete turn and pump the
handlebars lever slowly. As the fluid is ejected from the bleed
screw the level in the reservoir will fall. Take care that the level
does not drop too low whilst the operation continues, otherwise
air will re-enter the system, necessitating a fresh start.

5   Continue the pumping action with the lever until no further
air bubbles emerge from the end of the plastic pipe. Hold the
brake lever against the handlebars and tighten the caliper bleed
screw. Remove the plastic tube **after** the bleed screw is closed.

6   Check the brake action for sponginess, which usually denotes
there is still air in the system. If the action is spongy, continue
the bleeding operation in the same manner, until all traces of air
are removed.

7   When all traces of air have been removed from the system,
top up the reservoir and refit the diaphragm and cap or cover, as
appropriate. Check the entire system for leaks, and check also
that the brake system in general is functioning efficiently before
using the machine on the road.

8   Brake fluid drained from the system will almost certainly be
contaminated, either by foreign matter or more commonly by
the absorption of water from the air. All hydraulic fluids are to
some degree hygroscopic, that is, they are capable of drawing
water from the atmosphere, and thereby degrading their specifications. In view of this, and the relative cheapness of the fluid,
old fluid should always be discarded. Note: where twin disc brakes
are employed, the bleeding procedure should be repeated on the
other caliper.

7.2a Check condition of brake pipes and hoses ...

7.2b ... and ensure that all unions are tight

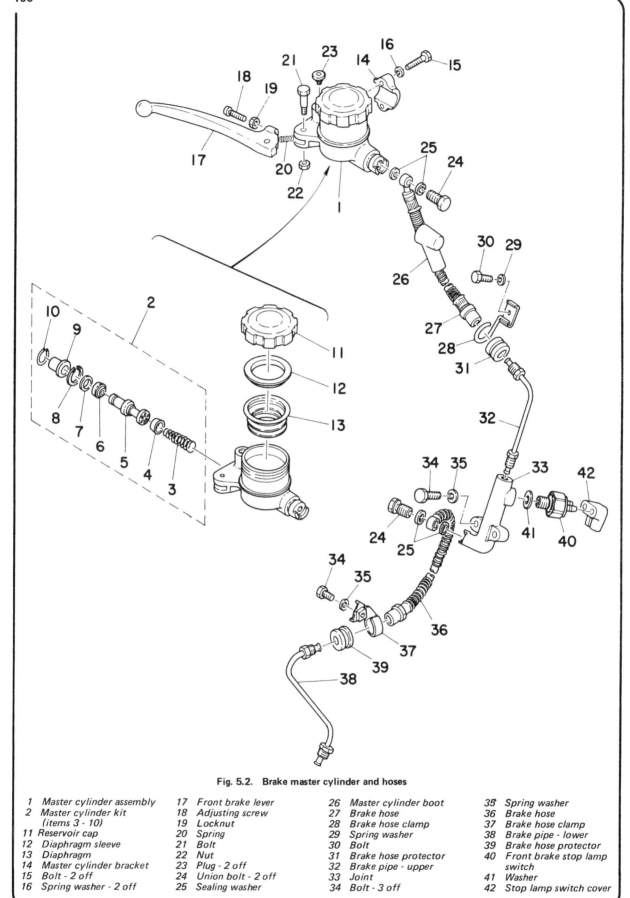

Fig. 5.2. Brake master cylinder and hoses

1 Master cylinder assembly	17 Front brake lever	26 Master cylinder boot	35 Spring washer
2 Master cylinder kit	18 Adjusting screw	27 Brake hose	36 Brake hose
(items 3 - 10)	19 Locknut	28 Brake hose clamp	37 Brake hose clamp
11 Reservoir cap	20 Spring	29 Spring washer	38 Brake pipe - lower
12 Diaphragm sleeve	21 Bolt	30 Bolt	39 Brake hose protector
13 Diaphragm	22 Nut	31 Brake hose protector	40 Front brake stop lamp
14 Master cylinder bracket	23 Plug - 2 off	32 Brake pipe - upper	switch
15 Bolt - 2 off	24 Union bolt - 2 off	33 Joint	41 Washer
16 Spring washer - 2 off	25 Sealing washer	34 Bolt - 3 off	42 Stop lamp switch cover

9 Front disc brake: recommended torque settings

Component	m-kg	ft-lbs
Disc mounting bolts	0.80–1.00	5.8–7.2
Caliper mounting bolts	4.00–5.00	28.9–36.1
Brake pipe union nut	1.30–1.80	9.4–13.0
Brake pipe/hose union nut	1.30–1.80	9.4–13.0
Banjo union nut at brake lever	1.50–2.00	10.8–14.5
Bleed screw	0.60–0.90	4.3–6.5

10 Front brake lever adjustment

The front brake lever should be checked periodically and adjusted at the stop bolt to give 13 mm - 20 mm (0.5 – 0.8 in) free play measured at the lever end.

11 Front wheel: removal, all models

1   With the machine supported on the centre stand, block the stand and crankcase to raise the wheel clear of the ground.
2   Detach the speedometer drive cable at the wheel by unscrewing the knurled gland nut which retains it. On drum brake wheels slacken the wheel spindle pinch bolt; disc brake models have a corresponding spindle clamp, which should be released by slackening the two bolts which retain it. Remove the front brake cable from drum brake models.
3   Remove the wheel spindle nut and split pin, and withdraw the spindle with the aid of a tommy bar. The wheel can now be lowered clear of the forks and put to one side.
4   On twin disc brake models. it will be found that the wheel rim and tyre will not pass between the two calipers. In this case, remove the mudguard mounting bolts and nuts from one of the lower fork legs. The leg can then be turned on the stanchion to swing the brake caliper out of the way. Alternatively, one of the calipers can be detached to obtain sufficient clearance.

8.3 Fit tube over bleed nipple before bleeding system

11.2 Speedometer cable is retained by gland nut

11.3 Wheel can be dropped clear of forks

11.4a Caliper can be detached to facilitate wheel removal

12 Front drum brake: examination and renovation

1   With the front wheel removed, as described in the preceding section, the twin leading shoe brake mechanism and backplate can be pulled free from the drum.

2   Examine the drum surface for signs of scoring or oil contamination. Both of these conditions will impair braking efficiency. Remove all traces of dust, preferably using a brass wire brush, taking care not to inhale any of it, as it is of an asbestos nature and consequently toxic. Remove oil or grease deposits, using a petrol soaked rag.

3   If deep scoring is evident, due to the linings having worn through to the shoe at some time, the drum must be skimmed on a lathe, or renewed. Whilst there are firms who will undertake to skim a drum whilst fitted to the wheel, it should be borne in mind that excessive skimming will change the radius of the drum in relation to the brake shoes therefore reducing the friction area until extensive bedding in has taken place. Also full adjustment of the shoes may not be possible. If in doubt about this point, the advice of one of the specialist engineering firms who undertake this work should be sought.

4   If fork oil or grease from the wheel bearings has badly contaminated the linings, they should be renewed. There is no satisfactory way of degreasing the lining material, which in any case is relat-

ively cheap to replace. It is a false economy to try to cut corners with brake components, the whole safety of both machine and rider being dependent on their condition.

5   The linings are bonded to the shoes, and the shoe must be renewed complete with the linings. This is accomplished by folding the shoes together until the spring tension is relaxed, and then lifting the shoes and springs off the brake plate. Fitting new shoes is a direct reversal of the above procedure.

6   Before refitting existing shoes, roughen the lining surface sufficiently to break the glaze which will have formed in use.

7   Examine the linkage which runs from the brake actuating lever to operate the second fulcrum. This may be removed after slackening the pinch bolts which retain it to the splines. The linkage may be further dismantled, if desired, by removing the circlips and clevises which retain the connecting rod to the actuating levers.

8   Push out the fulcrums from the brake plate. If there is corrosion on the fulcrum face or in its bore, this should be removed using wet or dry paper. Grease both fulcrums before installation. With the shoes, fulcrums and linkage in position on the brake plate, adjust the linkage by means of the turnbuckle and locknut on the connecting rod, so that the fulcrum faces are parallel to each other. This is an important point as both shoes must contact the drum simultaneously if full braking effect is to be obtained. Maladjustment between the two shoes can often cause mysterious juddering when the brake is applied lightly.

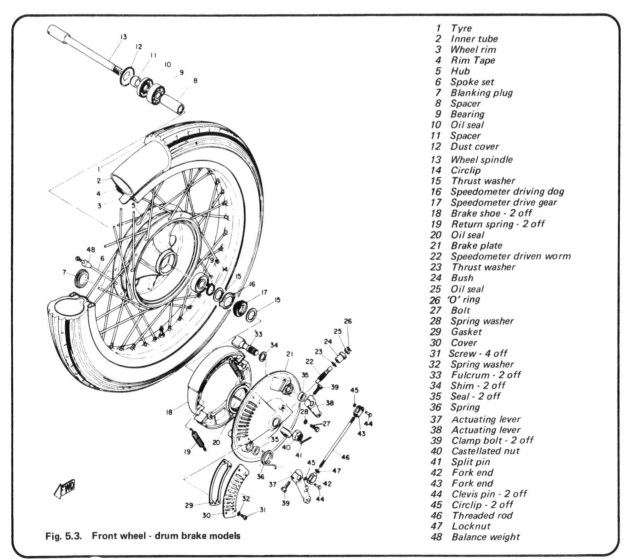

1	Tyre
2	Inner tube
3	Wheel rim
4	Rim Tape
5	Hub
6	Spoke set
7	Blanking plug
8	Spacer
9	Bearing
10	Oil seal
11	Spacer
12	Dust cover
13	Wheel spindle
14	Circlip
15	Thrust washer
16	Speedometer driving dog
17	Speedometer drive gear
18	Brake shoe - 2 off
19	Return spring - 2 off
20	Oil seal
21	Brake plate
22	Speedometer driven worm
23	Thrust washer
24	Bush
25	Oil seal
26	'O' ring
27	Bolt
28	Spring washer
29	Gasket
30	Cover
31	Screw - 4 off
32	Spring washer
33	Fulcrum - 2 off
34	Shim - 2 off
35	Seal - 2 off
36	Spring
37	Actuating lever
38	Actuating lever
39	Clamp bolt - 2 off
40	Castellated nut
41	Split pin
42	Fork end
43	Fork end
44	Clevis pin - 2 off
45	Circlip - 2 off
46	Threaded rod
47	Locknut
48	Balance weight

Fig. 5.3.   Front wheel - drum brake models

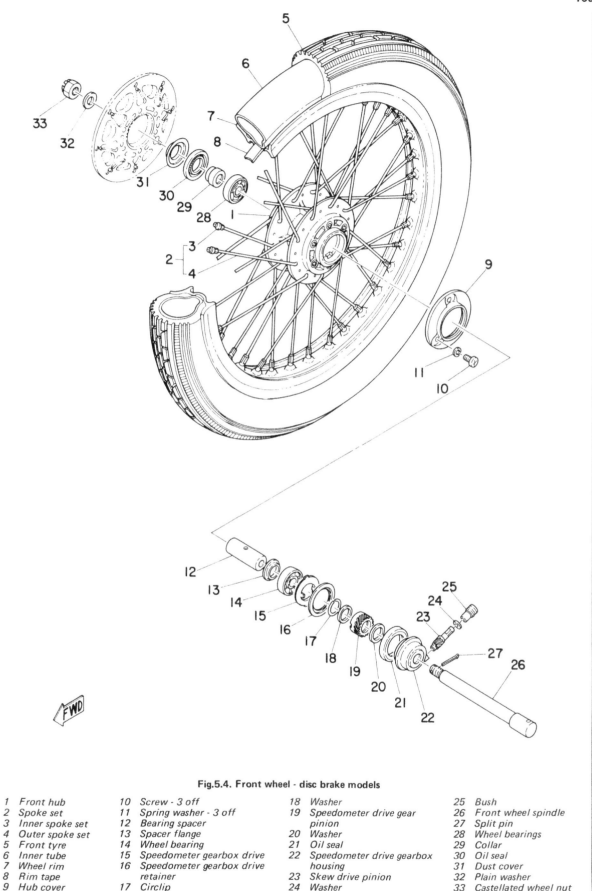

Fig.5.4. Front wheel - disc brake models

1	Front hub	10	Screw - 3 off	18	Washer	25 Bush
2	Spoke set	11	Spring washer - 3 off	19	Speedometer drive gear	26 Front wheel spindle
3	Inner spoke set	12	Bearing spacer		pinion	27 Split pin
4	Outer spoke set	13	Spacer flange	20	Washer	28 Wheel bearings
5	Front tyre	14	Wheel bearing	21	Oil seal	29 Collar
6	Inner tube	15	Speedometer gearbox drive	22	Speedometer drive gearbox	30 Oil seal
7	Wheel rim	16	Speedometer gearbox drive		housing	31 Dust cover
8	Rim tape		retainer	23	Skew drive pinion	32 Plain washer
9	Hub cover	17	Circlip	24	Washer	33 Castellated wheel nut

13 Rear wheel: removal and refitting

1   Make sure that the machine is supported securely on its centre stand. If a brake wear switch is fitted (photo 13.1a) disconnect its wiring. Remove the split pin and nut which retains the rear brake torque arm to the brake plate. Disconnect the brake operating rod. Release the rear wheel adjuster lock nuts, and slacken off the adjusters. Remove the split pin which locks the rear wheel spindle nut, and slacken and remove the nut.

2   Push the wheel forward so that the chain becomes slack, and prise off the spring link. Separate the drive chain and reassemble the joining link on one end of the chain to obviate any risk of loss.

3   With the chain detached, pull out the wheel spindle, catching the adjusters and spacer as they drop free. The wheel can now be disengaged from the swinging arm. Note that on XS650C models, two fillets are fitted in the fork ends and are each retained by a bolt. It is possible to remove these and withdraw the wheel, complete with spindle, if this is desired.

4   Refitting is a reversal of the removal procedure, noting that care must be taken to install the spacers on the correct side of the hub. When reconnecting the chain joining link ensure that the closed end of the spring link faces in the direction of normal chain travel. Adjustment of chain freeplay, wheel alignment and rear brake adjustment should be carried out as described in Sections 19 and 17 of this chapter. After tightening the spindle nut insert a new split pin in its hole and bend the legs over to secure.

5   Check the operation of the rear brake and stop light switch before riding the machine.

14 Wheel bearings and seals: examination and renovation

1   Access to the front wheel bearings may be made after removal of the wheel from the forks. Pull the speedometer gearbox out of the hub left-hand boss and remove the dust seal cover and wheel spacer from the hub right-hand side.

2   Lay the wheel on the ground with the disc side facing downward and with a special tool, in the form of a rod with a curved end, insert the curved end into the hole in the centre of the spacer separating the two wheel bearings. If the other end of the special tool is hit with a hammer, the right-hand bearing, bearing flange washer, and bearing spacer will be expelled from the hub.

3   Invert the wheel and drive out the left-hand bearing by inserting a drift of the appropriate size, through the hub. During the removal of either bearing it may be necessary to support the wheel across an open-ended box so that there is sufficient clearance for the bearing to be displaced completely from the hub.

4   Remove all the old grease from the hub and bearings, giving the latter a final wash in petrol. Check the bearings for signs of play or roughness when they are turned. If there is any doubt about the condition of a bearing, it should be renewed.

5   Before replacing the bearings, first pack the hub with new grease. Then drive the bearings back into position, not forgetting the distance piece that separates them. Take great care to ensure that the bearings enter the housings perfectly squarely otherwise the housing surface may be broached. Fit replacement oil seals and any dust covers or spacers that were also displaced during the original dismantling operation.

6   The rear seals and bearings can be dealt with in a similar manner to that described above. Reference should be made to the accompanying line drawings to aid correct reassembly;

13.1a Remove brake wear switch lead (where fitted)

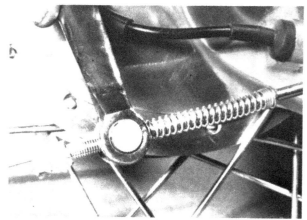

13.1b Disconnect rear brake operating rod

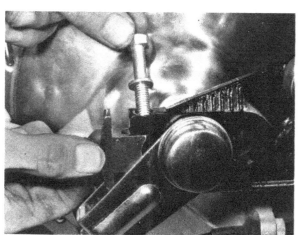

13.3a Remove fillets on XS650C models

13.3b Wheel can now be removed from the swinging arm

Tyre changing sequence - tubed tyres

 Deflate tyre. After pushing tyre beads away from rim flanges push tyre bead into well of rim at point opposite valve. Insert tyre lever adjacent to valve and work bead over edge of rim.

Use two levers to work bead over edge of rim. Note use of rim protectors

 Remove inner tube from tyre

When first bead is clear, remove tyre as shown

 When fitting, partially inflate inner tube and insert in tyre

Work first bead over rim and feed valve through hole in rim. Partially screw on retaining nut to hold valve in place.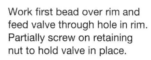

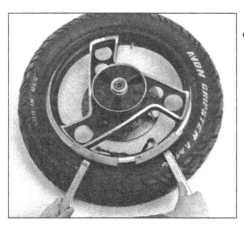

 Check that inner tube is positioned correctly and work second bead over rim using tyre levers. Start at a point opposite valve.

Work final area of bead over rim whilst pushing valve inwards to ensure that inner tube is not trapped

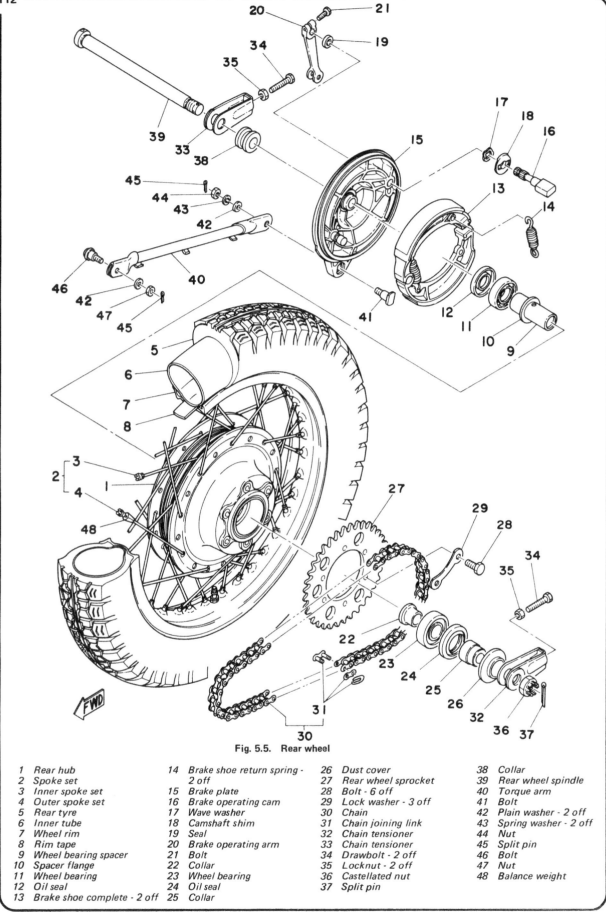

Fig. 5.5. Rear wheel

1	Rear hub	14	Brake shoe return spring -	26	Dust cover	38	Collar
2	Spoke set		2 off	27	Rear wheel sprocket	39	Rear wheel spindle
3	Inner spoke set	15	Brake plate	28	Bolt - 6 off	40	Torque arm
4	Outer spoke set	16	Brake operating cam	29	Lock washer - 3 off	41	Bolt
5	Rear tyre	17	Wave washer	30	Chain	42	Plain washer - 2 off
6	Inner tube	18	Camshaft shim	31	Chain joining link	43	Spring washer - 2 off
7	Wheel rim	19	Seal	32	Chain tensioner	44	Nut
8	Rim tape	20	Brake operating arm	33	Chain tensioner	45	Split pin
9	Wheel bearing spacer	21	Bolt	34	Drawbolt - 2 off	46	Bolt
10	Spacer flange	22	Collar	35	Locknut - 2 off	47	Nut
11	Wheel bearing	23	Wheel bearing	36	Castellated nut	48	Balance weight
12	Oil seal	24	Oil seal	37	Split pin		
13	Brake shoe complete - 2 off	25	Collar				

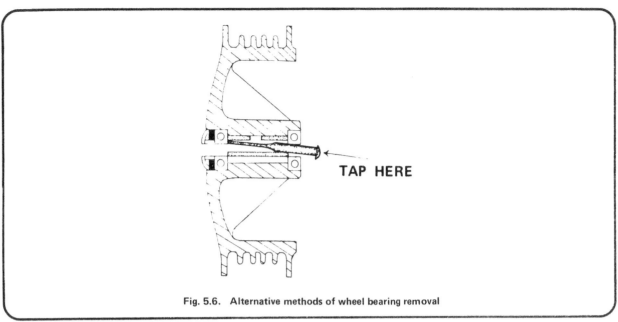

Fig. 5.6. Alternative methods of wheel bearing removal

14.5a Note position of spacers ...

14.5b ... and oil seals when removing bearings

14.5c Seals must be renewed **every** time

14.5d Bearings can be driven out of the hub

14.5e Note spacer tube between bearings

15 Speedometer drive gearbox: examination and lubrication

1   The speedometer drive is taken from a gearbox mounted on
the front wheel of all models. On drum brake models, the drive
gear is supported in the brake plate casting, as is the driven shaft.
The drive gear can be removed from the inside of the brake plate
together with the driving dog, after removing the circlip which
retains it. The driven shaft can be removed from the outside of
the brake plate, by unscrewing its retaining gland nut. Unless
excessively worn due to lack of lubricant, which will necessitate renewal, little can be done by way of maintenance other
than thorough cleaning and relubrication of the components and
their housing.
2   Maintenance of the disc brake type of gearbox should be
approached in the same manner; in this case, the unit being a
self-contained assembly which is a push fit in the front hub.
The unit should be greased as a matter of course, each time the
front wheel is removed.

16   Rear drum brake: examination and renovation

1   Remove the rear wheel as described in Section 13 and lift
out the drum brake for examination. Although of the single
leading shoe design, the examination and renovation procedure
for all components is the same as detailed for the front brake in
Section 12 of this chapter. Any comments relating to the
operating linkage of the twin leading shoe front brake can be
disregarded.

15.2a Drive gearbox should be greased occasionally

2   On models which have the integral brake shoe wear
indicator, check that the switch contact is clean and operates
freely. If necessary, unscrew the switch from the brake
backplate and spray with switch contact cleaner.

17 Adjusting the rear brake

1   If the adjustment of the rear brake is correct, the rear brake
pedal will have about 25 mm (1 inch) free play before the brake
commences to operate
2   The length of travel is controlled by the adjuster at the end of
the brake operating rod, close to the brake operating arm. If the
nut is turned clockwise, the amount of travel is reduced and
vice-versa. Always check that the brake is not binding after adjustments have been made.
3   Note that it may be necessary to re-adjust the height of the
stop lamp switch if the pedal height has been changed to any
marked extent. The switch is located immediately below the
right-hand side cover that carries the capacity symbol of the
model. The body of the switch is threaded, so that it can be
raised or lowered, after the locknuts have been slackened. If the
stop lamp lights too soon, the switch should be lowered, and
vice-versa.

18 Rear wheel sprocket: removal, examination and replacement

1   The rear wheel sprocket assembly can be removed as a separate unit after the rear wheel has been separated and detached
from the frame as described in Section 13 of this Chapter. It is
retained by six bolts which are locked with tab washers.
2   Check the condition of the sprocket teeth. If they are
hooked, chipped or badly worn, the sprocket must be renewed.
3   It is considered bad practice to renew one sprocket on its
own. The final drive sprockets should always be renewed as a
pair and a new chain fitted, otherwise rapid wear will necessitate even earlier renewal on the next occasion.
4   It should be noted that regular chain maintenance will greatly
prolong the life of the chain and its sprockets, whilst a worn or
maladjusted chain will quickly destroy the sprocket.

19 Final drive chain: examination and lubrication

1   The final drive chain is fully exposed, with only a light chainguard over the top run. Periodically the tension will need to be
adjusted, to compensate for wear. This is accomplished by
placing the machine on the centre stand and slackening the
wheel nut on the left-hand side of the rear wheel so that the
wheel can be drawn backward by means of the drawbolt adjusters in the fork ends. The rear brake torque arm bolt must also be
slackened during this operation.

15.2b Note the locating slot in the gearbox

2 The chain is in correct tension if there is approximately 20 mm (¾ in) slack in the middle of the lower run. Always check when the chain is at its tightest point as a chain rarely wears evenly during service.

3 Always adjust the drawbolts an equal amount in order to preserve wheel alignment. The fork ends are clearly marked with a series of horizontal lines above the adjusters, to provide a simple visual check. If desired, wheel alignment can be checked by running a plank of wood parallel to the machine, so that it touches the sides of the rear tyre. If wheel alignment is correct, the plank will be equidistant from each side of the front wheel tyre, when tested on both sides of the rear wheel. It will not touch the front wheel tyre because this tyre is of smaller cross section. See accompanying diagram.

4 Do not run the chain overtight to compensate for uneven wear. A tight chain will place undue stress on the gearbox and rear wheel bearings, leading to their early failure. It will also absorb a surprising amount of power.

5 After a period of running, the chain will require lubrication. Lack of oil will greatly accelerate the rate of wear of both the chain and the sprockets and will lead to harsh transmission. The application of engine oil will act as a temporary expedient, but it is preferable to remove the chain and clean it in a paraffin bath before it is immersed in a molten lubricant such as Linklife or Chainguard. These lubricants achieve better penetration of the chain links and rollers and are less likely to be thrown off when the chain is in motion.

6 To check whether the chain is due for replacement, lay it lengthwise in a straight line and compress it endwise so that all the play is taken up. Anchor one end and measure the length. Now pull the chain with one end anchored firmly, so that the chain is fully extended by the amount of play in the opposite direction. If there is a difference of more than ¼ inch per foot in the two measurements, the chain should be renewed in conjunction with the sprockets. Note that this check should be made **after** the chain has been washed out, but **before** any lubricant is applied, otherwise the lubricant may take up some of the play.

7 When replacing the chain, make sure that the spring link is seated correctly, with the closed end facing the direction of travel.

19.3a Note that adjusters should be parallel and ...

19.3b ...check that location marks correspond

19.7 Closed end of link must face direction of travel

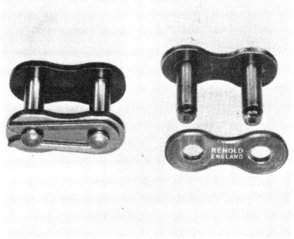

19.8 An equivalent British-made chain is available

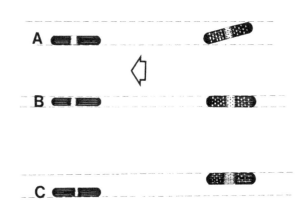

Fig. 5.7. Method of checking wheel alignment

A & C – Incorrect   B – Correct

20 Chain and sprocket conversions

1   The chain fitted as standard is of Japanese manufacture. When renewal becomes necessary, it should be noted that a Renold equivalent, of British manufacture, is available. When purchasing a replacement, take along the old chain as a pattern or, if known, a note of the size and number of pitches.

2   Despite widespread belief to the contrary, there is rarely any advantage to be gained from varying sprocket sizes from those specified by the manufacturer. A larger gearbox sprocket by no means guarantees a higher maximum speed; usually it converts top gear into little more than an overdrive and may lead to a situation where the machine is faster in the next lower gear.

21 Tyres - removal and replacement

1   At some time or other the need will arise to remove and replace the tyres, either as a result of a puncture or because a renewal is required to offset wear. To the inexperienced tyre changing represents a formidable task yet if a few simple rules are observed and the technique learned, the whole operation is surprisingly simple.

2   To remove the tyre from either wheel, first detach the wheel from the machine by following the procedure in Chapter 5 Sections 11 or 13 depending on whether the front or the rear wheel is involved. Deflate the tyre by removing the valve insert and when it is fully deflated, push the bead of the tyre away from the wheel rim on both sides so that the bead enters the centre well of the rim. Remove the locking cap and push the tyre valve into the tyre itself.

3   Insert a tyre lever close to the valve and lever the edge of the tyre over the outside of the wheel rim, Very little force should be necessary; if resistance is encountered it is probably due to the fact that the tyre beads have not entered the well of the wheel rim all the way round the tyre.

4   Once the tyre has been edged over the wheel rim, it is easy to work around the wheel rim so that the tyre is completely free on one side. At this stage, the inner tube can be removed.

5   Working from the other side of the wheel, ease the outer edge of the tyre over the outside of the wheel rim which is furthest away. Continue to work around the rim until the tyre is free completely from the rim.

6   If a puncture has necessitated the removal of the tyre, re-inflate the inner tube and immerse it in a bowl of water to trace the source of the leak. Mark its position and deflate the tube. Dry the tube and clean the area around the puncture with a petrol soaked rag. When the surface has dried, apply the rubber solution and allow this to dry before removing the backing from the patch and applying the patch to the surface.

7   It is best to use a patch of the self-vulcanising type which will form a very permanent repair. Note that it may be necessary to remove a protective covering from the top surface of the patch. after it has sealed in position. Inner tubes made from synthetic rubber may require a special type of patch and adhesive if a satisfactory bond is to be achieved.

8   Before refitting the tyre, check the inside to make sure that the agent which caused the puncture is not trapped. Check the outside of the tyre, particularly the tread area, to make sure nothing is trapped that may cause a further puncture.

9   If the inner tube has been patched on a number of past occasions, or if there is a tear or large hole, it is preferable to discard it and fit a new one. Sudden deflation may cause an accident, particularly if it occurs with the front wheel.

10 To refit the tyre, inflate the inner tube sufficiently for it to assume a circular shape but only just. Then push it into the tyre so that it is enclosed completely. Lay the tyre on the wheel at an angle and insert the valve through the rim tape and the hole in the wheel rim. Attach the locking cap on the first few threads, sufficient to hold the valve captive in its correct location.

11 Starting at the point furthest from the valve, push the tyre bead over the edge of the wheel rim until it is located in the central well. Continue to work around the tyre in this fashion until the whole of one side of the tyre is on the rim. It may be necessary to use a tyre lever during the final stages.

12 Make sure that there is no pull on the tyre valve and again commencing with the area furthest from the valve, ease the other bead of the tyre over the edge of the rim. Finish with the area close to the valve, pushing the valve up into the tyre until the locking cap touches the rim. This will ensure the inner tube is not trapped when the last section of the bead is edged over the rim with a tyre lever.

13 Check that the inner tube is not trapped at any point. Re-inflate the inner tube and check that the tyre is seating correctly around the wheel rim. There should be a thin rib moulded around the wall of the tyre on both sides which should be equidistant from the wheel rim at all points. If the tyre is unevenly located on the rim, try bouncing the wheel when the tyre is at the recommended pressure. It is probable that one of the beads has not pulled clear of the centre well.

14 Always run the tyres at the recommended pressures and never under or over-inflate. The correct pressures for solo use are given in the Specifications Section of this Chapter. If a pillion passenger is carried, increase the rear tyre pressure only by approximately 4 psi.

15 Tyre replacement is aided by dusting the side walls, particularly in the vicinity of the beads, with a liberal coating of French Chalk. Washing up liquid can also be used to good effect, but this has the disadvantage of causing the inner surfaces of the wheel rim to rust.

16 Never replace the inner tube and tyre without the rim tape in position. If this precaution is overlooked there is a good chance of the ends of the spoke nipples chafing the inner tube and causing a crop of punctures.

17 Never fit a tyre which has a damaged tread or side walls. Apart from the legal aspects, there is a very great risk of a blow-out, which can have serious consequences;

22 Fault diagnosis

Symptom	Cause	Remedy
Handlebars oscillate at low speeds	Buckle or flat in wheel rim, most probably front wheel	Check rim alignment by spinning wheel Correct by retensioning spokes or rebuilding on new rim.
	Tyre not straight on rim	Check tyre alignment
Machine lacks power and accelerates poorly	Rear brake binding	Warm brake drum provides best evidence. Re-adjust brake
Rear brake grabs when applied gently	Ends of brake shoes not chamfered Elliptical brake drum	Chamfer with file Lightly skim in lathe (specialist attention required).
Front brake feels spongy	Air in hydraulic system	Bleed brake
Brake pull-off sluggish	Brake cam binding in housing (drum brake) Weak brake shoe springs (drum brake) Sticking pistons in brake caliper	Free and grease Renew if springs have not become displaced Overhaul caliper unit
Harsh transmission	Worn or badly adjusted final drive chain Hooked or badly worn sprockets Worn or deteriorating cush drive rubbers	Adjust or renew as necessary Renew as a pair Renew rubbers

Chapter 6 Electrical system

Refer to Chapter 7 for details of 1977 to 1983 models

Contents

Specifications

Battery

Make	Yuasa	or GS
Type	YB 14L	12N12-4A-1
Voltage	12 Volts	12 volts
Capacity	14 Ah	12 Ah
Earth	Negative	Negative

Alternator

Make	Hitachi
Model	LD 115-02 or LD 115
Type	Brush type
Output	11 amps @ 2,000 rpm

Starter motor

Make	Hitachi
Model	S108-35
Output	0.5 Kw

Regulator

Make	Hitachi	or Hitachi
Type	TLIZ-80	or TLIZ-49
No-load output	14.5 ± 0.5 V @3000 rpm	14.5 ± 0.5 V @ 2000 rpm

Rectifier

Make	Hitachi
Type	SD6D-9

Bulbs

Headlamp	12 V 50/40W sealed beam unit
Stop/tail light	12 V 8/23W
Indicators	12 V 27W
Pilot light	12 V 5W
Neutral light	12 V 3W
Indicator warning light	12 V 3W
Main beam warning light	12 V 3W
Instrument illumination lights	12 V 3W
Brake wear warning light	12 V 3W
Brake light warning light	12 V 3W

1 General description

The Yamaha 650 twins are fitted with a 12 volt electrical system powered by an alternator mounted on the left-hand side of the crankshaft. Current from the alternator is passed through a full-wave silicon rectifier, where it is converted from alternating current (ac) to direct current (dc). The rectified current is then routed to the voltage regulator, which matches the generator output to the varying requirements of the electrical system. Excess power is used to charge the battery when required, and to maintain the alternator output at the appropriate level.

2 Checking the charging system output

1   It is possible to perform checks on the alternator, and the electrical system in general, using a test meter of the multimeter type. It is assumed that the owner is reasonably conversant with its use and that it is appreciated that unwitting reversal or open circuiting of some connections may irreparably damage some components of the electrical system. It is recommended therefore, that the owner should entrust any tests or repairs to a Yamaha Service Agent or a qualified Auto-electrician if in any doubt about the correct procedure.
2   Disconnect the battery positive lead at the main fuse. Set the multimeter on the 0-20 volts dc range, and connect the red (+) probe to the regulator side of the separated battery lead. The black (–) probe should be connected to a convenient earth point. With the engine running at approximately 2500 rpm, the meter should show a reading of 14·5 - 15 volts.
3   Should the readings differ noticeably from those given, and the alternator and regulator are known to be in good condition, it is possible to adjust the regulator by means of an adjuster screw. Remove the right-hand side panel to expose the regulator unit. Remove the regulator cover screws and lift the cover away. The adjuster screw is located on the front face of the unit and may be screwed inwards to raise the voltage, and out to lower it.

3 Voltage regulator: testing

1   Should the operation of the regulator be suspect, it can be tested as described below, having first checked it visually for loose or frayed wires or burnt contacts. Should the contacts appear burnt or pitted, they can be dressed in a similar manner to ignition contact breaker points, using a fine swiss file. Having established that the unit is mechanically sound, the following procedure should be adopted:
2   Set the multimeter to the resistance scale, and disconnect the multi-way connector from the regulator. Connect one probe to the black lead, and the other to the base of the regulator. The meter should indicate no resistance. If this is not the case, the black lead will probably be frayed or broken and should be renewed.
3   Connect one of the meter probes to the brown lead at the connector block, and the other to the green lead. In its normal rest position, with the central and top contacts closed, the meter should indicate no resistance. If a reading of 2 ohms or higher is indicated, check for broken or damaged wires or a badly soldered connection. Alternatively, check that the contacts are not burnt or pitted. If none of these steps cause any improvement, it is likely that the unit will need renewal. Take it to a Yamaha Service Agent for verification.
4   With the meter connected as described in paragraph 3, depress the central contact of the regulator so that it is midway between the top and bottom fixed contacts. In this position, a reading of 9 - 10 ohms is to be expected. If this is not the case, the 10 ohm resistor fitted to the unit is defective and should be renewed, preferably by a Yamaha Service Agent or a qualified Auto-electrician.

5   Retaining the meter connections, press the central contact down so that it touches the bottom fixed contact. The meter should indicate 7 - 8 ohms resistance. If this is not the case, ensure that the contact faces are in good condition and repeat the test.
6   Connect one of the meter probes to the brown lead at the connector, and the other to the black lead. With the central contact in its normal position, against the top contact, a resistance of 36 - 38 ohms should be indicated. Should any of the above tests fail to give the correct readings, and the various leads and soldered joints are in good condition, it may be assumed that the regulator is defective and it should be renewed or repaired. This work should be entrusted to someone qualified in this field, who will have the requisite experience to ensure that it is carried out correctly. If the unit conforms to the various resistance values, it is safe to assume that the fault lies elsewhere in the charging system.

3.1 Regulator is mounted beneath left-hand side panel

4 Rectifier: testing

1   As mentioned earlier, the rectifier converts ac current from the alternator, into dc current for the charging circuit. The type used on the Yamaha 650 Twins is a full-wave rectifier which converts the whole of the wavelength from the alternator to dc. The rectifier contains six diodes which act as one-way valves, allowing current to pass in one direction only. These can be checked very simply, using a multimeter set in the resistance mode:
2   Trace the wiring from the rectifier, which is mounted beneath the battery carrier, to its multi-way connector block. Separate the connector to expose its terminals. Attach the black meter probe (-) to the black connector terminal and touch the red meter probe (+) to each of the white leads in turn. A low resistance can be expected, around 9-10 ohms at the most, indicating that current can flow in that direction. If the meter probes are reversed, and the test is repeated, a very high resistance should be indicated by the meter, showing that the current flow is being stopped by the diode.
3   Using the red connector terminal instead of the black, the same test should be conducted. The results should be identical to those given above. Should the tests show that one or more of the diodes have broken down, and are allowing current to pass in both directions, the rectifier must be renewed. Normally, this will occur only when the rectifier has been overloaded as a result of reversed connections or by a shorted cable.

5 Alternator: testing

1　Should the alternator output be suspect, remove the inspection cover on the left-hand outer casing to gain access to the stator. Examine the brushes for signs of wear or damage. If necessary, remove the brush holder retaining screws and remove the holder and brushes for close examination and measurement. The nominal brush length is 14.5 mm (0.57 in). If the brushes are worn below 7.0 mm (0.28 in) they should be renewed. If within limits, it is possible to clean the brush ends, using a fine file or abrasive paper, taking care not to remove too much material and keeping the ends square.

2　Visually check the rotor slip-rings upon which the brushes bear, for signs of corrosion or contamination. It is permissible to remove any accumulation of glaze, using very fine polishing paper, such as Crokus paper, which is specifically designed for cleaning electrical contact surfaces. It is normally possible to do this via the windows in the stator.

3　Check the leads from the brushes to the multi-way connector, using a multimeter set on resistance. Note that any resistance in these leads will have a marked effect on the output of the alternator, and in consequence they must be renewed.

4　The rotor windings can be checked without removing the stator, by passing the multimeter probes through the brush holder aperture. The resistance, measured by placing a probe on each slip ring, should be between 5-7 ohms. The insulation

between the rotor core and each slip ring can be measured by placing one probe on the central retaining nut, and the other on each slip ring in turn. Should anything other than sound insulation be found, it is likely that the windings are broken or shorted internally, or shorting to the rotor core itself. If this is the case, it will be necessary to renew the rotor, as there is no practicable way of repairing it.

6 Alternator: removing and replacing the stator and rotor

1　In order to gain access to the alternator for removal purposes, it is necessary to detach the left-hand outer casing. The stator is retained by two long screws, and is located by a small dowel pin at the bottom of the crankcase supporting lugs. Take care not to lose this as the stator is lifted off. If the stator is to be renewed or removed for testing, the wiring clip should be opened to release the cables and the multi-way connector separated.

2　The rotor is keyed to a taper on the crankshaft and is retained by a central nut. It is advised that the manufacturers' rotor extractor is used whenever possible, as this bears on the rotor boss, obviating any risk of damage. In an emergency, a three legged puller can be used, bearing against the inner pole pieces. Note that **on no account** must a puller be used on the outer pole pieces, as this will almost certainly ruin the rotor by pulling the halves apart.

4.2 Rectifier is bolted to underside of battery tray

5.1a Brush holder can be detached to check brush condition

5.1b Note that only five of the screws retain the holder

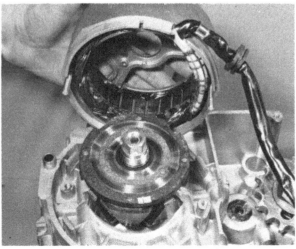

6.1 Remove alternator stator, noting locating pin

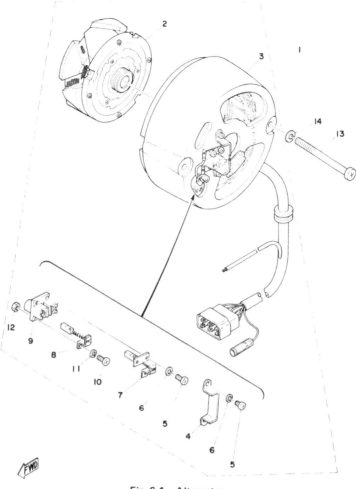

Fig. 6.1. Alternator

1	Alternator assembly	5	Screw - 2 off	9	Brush holder
2	Rotor assembly	6	Spring washer - 2 off	10	Screw - 4 off
3	Stator assembly	7	Brush assembly	11	Spring washer - 4 off
4	Shroud	8	Brush assembly	12	Nut

13	Screw - 2 off
14	Spring washer - 2 off

6.2 Use puller very carefully when removing rotor

7 Battery: examination and maintenance

1   A Yuasa or GS battery of either 12 or 14 Ah is fitted to the
Yamaha 650 twins.
2   The transparent plastic case of the battery permits the upper
and lower levels of the electrolyte to be observed without disturbing the battery by removing the left-hand side cover.
Maintenance is normally limited to keeping the electrolyte level
between the prescribed upper and lower limits and making sure
that the vent tube is not blocked. The lead plates and their
separators are also visable through the transparent case, a further
guide to the general condition of the battery.
3   Unless acid is spilt, as may occur if the machine falls over, the
electrolyte should always be topped up with distilled water to
restore the correct level. If acid is spilt onto any part of the
machine, it should be neutralised with an alkali such as washing
soda or baking powder and washed away with plenty of water,
otherwise serious corrosion will occur. Top up with sulphuric
acid of the correct specific gravity (1.260 to 1.280) only when
spillage has occured. Check that the vent pipe is well clear of the
frame or any of the other cycle parts.

4 It is seldom practicable to repair a cracked battery case because the acid present in the joint will prevent the formation of an effective seal. It is always best to renew a cracked battery, especially in view of the corrosion which will be caused if the acid continues to leak.

5 If the machine is not used for a period, it is advisable to remove the battery and give it a refresher charge every six weeks or so from a battery charger. If the battery is permitted to discharge completely, the plates will sulphate and render the battery useless.

6 Occasionally, check the condition of the battery terminals to ensure that corrosion is not taking place and that the electrical connections are tight. If corrosion has occurred, it should be cleaned away by scraping with a knife and then using emery cloth to remove the final traces. Remake the electrical connections whilst the joint is still clean, then smear the assembly with petroleum jelly (NOT grease) to prevent recurrence of the corrosion. Badly corroded connections can have a high electrical resistance and may give the impression of a complete battery failure.

8 Battery: charging procedure

1 Since the ignition system is dependent on the battery for its operation, if the battery discharges completely it must be removed and recharged before the machine can be used. A battery charger is necessary for this purpose.

2 The normal charge rate is 2 amps for about 6 hours for a 12 amp hour battery. A more rapid charge at a higher rate can be given in an emergency, but this should be avoided if at all possible because it will shorten the useful working life of the battery. Always ensure the battery is topped up before charging.

3 When the battery is replaced on the machine, make sure that it is protected by the rubber pads in the battery compartment, which will help damp out the undesirable effects of vibration. Do not reverse the connections to the battery, or the sillicon rectifier may be damaged.

9 Fuse - location and replacement

1 A fuse is incorporated in the electrical system to give protection to the bulbs etc in the case of sudden overload caused through a short circut. It is located at the right-hand side of the battery in a plastic container. The fuse is rated at 15 amps.

2 If a fuse blows, it should not be replaced until a check has shown whether a short circuit has occurred. This will involve checking the wiring circuit to find and rectify the fault. If this procedure is not adopted, the replacement, which may be the only spare, may blow immediately on reconnection.

3 When a fuse blows whilst the machine is running and no spare is available, a 'get you home' remedy is to remove the blown fuse and wrap it in silver paper before replacing it in the fuse holder. The silver paper will restore the electric circuit by bridging the blown wire inside the fuse. This expedient should never be used as a permanent measure otherwise more serious damage will be done to the electrical system.

4 The doctored fuse must be replaced as soon as possible to restore full circuit protection. It is an emergency repair only.

10 Electric starter - general description

1 An electric starter is fitted to all models from the XS 2 onwards. In each case, the motor is mounted at the back of the crankcase, on the underside of the unit, driving the engine through reduction gears. On the XS 2 and early TX 650 models, the starter switch was linked with a decompressor lever mounted on the handlebars. As the lever is pulled, a fulcrum on the left-hand exhaust valve inspection cover partially opens the valve. This decompresses the left-hand cylinder, effectively halving the amount of resistance offered to the starter motor. As the decompressor operates, the starter switch is closed, bringing the starter solenoid into operation. The latter switches the heavy current required to the starter motor which turns the decompressed engine until it starts. As soon as the engine begins to run, an override circuit switches off the starter motor to prevent damage through over-revving. The decompressor can then be released and the engine run normally.

2 On later models, the decompressor device has been dispensed with, the starter motor cranking the engine at full compression. The starter switch on these models takes the form of a push button incorporated in the right-hand switch unit. The override circuit has been discarded on these models.

7.2 Battery electrolyte level is visible through case

9.1 Note spare fuse contained in fuse holder

11 Starter motor: removal and dismantling

1   Before any attempt is made to remove the starter motor, the engine oil must be drained into a container of at least 3.2 litres (5.6 Imp pints, 3.4 US qts) depending on the model. If this operation is overlooked, it will do so of its own accord as the starter motor is withdrawn from the reduction gear housing.
2   Detach the starter motor cable. It will probably be found easier to disconnect it at the solenoid end and feed the cable through as the motor is withdrawn, as the starter motor terminal is very close to the frame. The motor is retained by four bolts which pass through brackets into the crankcase. The motor can be withdrawn after these have been removed.
3   Remove the two long bolts which hold the end covers to the motor. Note that the remaining two screws on the closed end cover retain the brush assembly. Remove the end covers carefully to avoid any risk of damage to the oil seal or the two 'O' rings. The armature can now be withdrawn from the casing.

12 Starter motor: examination, checking and renovation

1   With the starter motor removed and dismantled, the carbon brushes should be checked for wear and condition. The nominal length of the brushes is 14.5 mm (0.6in). If worn to 7.0 mm (0.28in) or less, they should be renewed. Check that the working face of each brush is clean. A solvent, such as petrol can be used to remove any deposits. If necessary, use a fine file to restore the rubbing face, but beware of removing too much material. The brushes should slide freely in their holders but without excessive side to side movement.
2   When fitted to the holders, the brushes should bear upon the commutator with a pressure of 600-880 grams (21-31 oz.). In practice, it is difficult to contrive a means of measuring this without the correct equipment, and it will normally suffice to ensure that the brushes are in firm contact with the commutator segments.
3   Check the armature visually before cleaning, especially if it has malfunctioned suddenly. Look for traces of molten solder at the commutator end, and at the same time check the inside of the casing. Should blobs of solder be in evidence, then it is likely that the motor has been over-loaded and has burnt out. This can be confirmed by taking the armature to an Auto-Electrician who may possibly be able to repair it. In the case of internal shorts, it is usual to renew the armature in view of the work involved in repairing and rewinding the unit.
4   Examine the commutator segments for wear and scoring. The latter is normally caused by lack of maintenance leading to tracking, and consequently localised burning of the segment

faces. If merely dirty, the segments may be burnished by wrapping fine glasspaper and turning the armature until a bright finish is obtained, on no account use emery cloth or wet or dry paper, as the abrasive particles can become embedded in the soft copper. Wipe the commutator with a rag soaked in methylated spirits to remove any grease. In cases of light wear, it may be possible to skim the commutator in a lathe to restore the surface, otherwise the armature must be renewed.
5   The commutator segments are insulated by a layer of mica which is designed to be 0.5 – 0.8 mm (0.02 – 0.03 in) below the surface of the segments. This air gap must be cleaned out, and if necessary re-cut to the correct depth. A broken hacksaw blade can be ground down to make a useful tool for the purpose.
6   The commutator segments should be tested using a multimeter set on resistance. Each segment should be tested by placing one of the meter probes on any one segment and touching each of the other segments in turn with the other probe. A complete circuit (no resistance) should be indicated. The insulation between the armature core and each commutator segment should be checked in a similar manner. The meter should indicate at least 3 megohms resistance. It should be noted that a tendency for a starter motor to fail spasmodically can often be attributed to a dead or a shorting segment in the commutator.
7   The field coils, which are mounted on the inside of the casing, should also be checked for resistance. The value should be between 0.045 and 0.055 ohms @ 20° C. Check also that there is a resistance of at least 0.1 megohms between the field coils and earth.

11.2 Starter motor is retained by four bolts

11.3 Note that two of the screws retain brush holder

12.1 Check condition of brushes and commutator

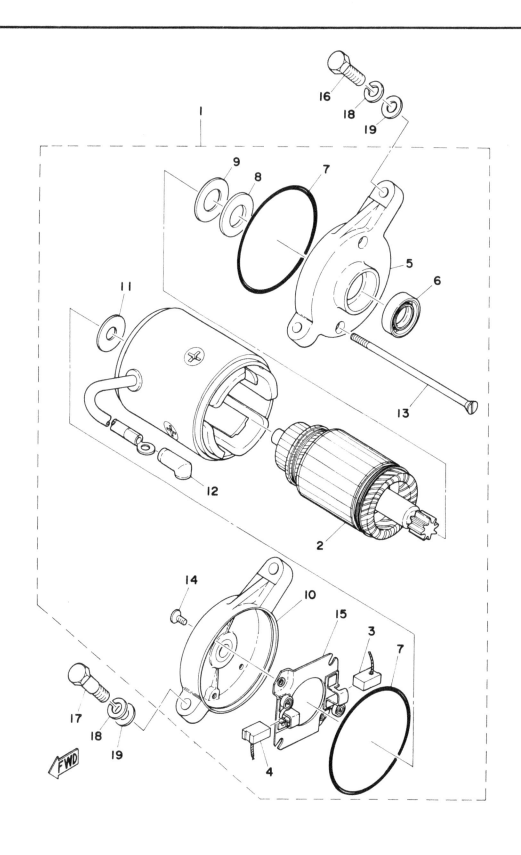

Fig. 6.2 Starter motor

1	Starter Motor Assembly	6	Oil seal	11	Thrust washer - 2 off
2	Armature	7	O ring - 2 off	12	Cap
3	Brush	8	Shim	13	Screw - 2 off
4	Brush	9	Thrust washer	14	Screw - 2 off
5	Starter motor cover	10	Starter motor cover	15	Brush holder

16 Bolt - 2 off
17 Bolt
18 Spring washer
19 Plain washer

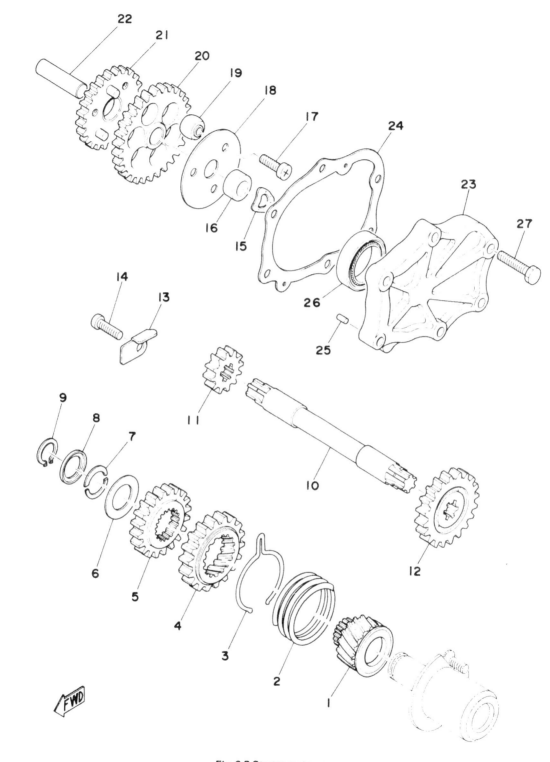

Fig. 6.3 Starter motor gears

1 Pinion	8 Retainer	15 Wave washer	22 Shaft
2 Spring	9 Circlip	16 Collar	23 Cover
3 Friction clip	10 Shaft	17 Screw - 3 off	24 Gasket
4 Pinion (25T)	11 Pinion (14T)	18 Plate	25 Dowel pin - 2 off
5 Pinion (23T)	12 Pinion (24T)	19 Shock absorber rubber - 6 off	26 Oil seal
6 Thrust washer	13 Stopper plate	20 Idler gear (36T)	27 Bolt - 6 off
7 Half clip - 2 off	14 Screw	21 Idler gear (26T)	

13 Starter motor: dimensions and tolerances

Field coil resistance	:	0.045−0.055 ohms @ 20°C
Commutator diameter	:	33 mm (1.299 in) nominal
Minimum diameter	:	32 mm (1.259 in)
Undercut (nominal)	:	0.5 − 0.8 mm(0.02−0.03 in)
Undercut (minimum)	:	0.2 mm (0.008 in)
Maximum runout	:	± 0.15 mm (0.006 in)

14 Starter solenoid switch: function and location

1   The starter motor switch is designed to work on the electro-magnetic principle. When the starter motor button is depressed, current from the battery passes through windings in the switch solenoid and generates an electro-magnetic force which causes a set of contact points to close. Immediately the points close, the starter motor is energised and a very heavy current is drawn from the battery.

2   This arrangement is used for at least two reasons. Firstly, the starter motor current is drawn only when the button is depressed and is cut off again when the pressure on the button is released. This ensures minimum drainage on the battery. Secondly, if the battery is in a low state of charge, there will not be sufficient current to cause the solenoid contacts to close. In consequence, it is not possible to place an excessive drain on the battery which, in some circumstances, can cause the plates to overheat and shed their coatings. If the starter will not operate, first suspect a discharged battery. This can be checked by trying the horn or switching on the lights. If this check shows the battery to be in good shape, suspect the starter solenoid which should come into action with a pronounced click. It is located behind the right-hand side panel and can be identified by the heavy duty starter cable connected to it. It is not possible to effect a satisfactory repair if the solenoid malfunctions; it must be renewed.

15 Headlamp: replacing the bulbs and adjusting beam height

1   In order to gain access to the headlamp bulbs it is necessary to first remove the rim, complete with the reflector and headlamp glass. The rim is retained by two screws which pass through the headlamp shell just below the two headlamp mounting bolts.

2   Headlamp units fitted to the 650 Twins are either of the sealed beam type or of the separate bulb type. With the sealed beam unit, if either of the headlamp filaments blow then the complete unit should be replaced. It will be found that a reflector that accepts a pilot bulb is fitted to all models delivered to countries or states where parking lights are a statutory requirement. The pilot bulb is held in the bulb holder by a bayonet fixing.

3   Beam height on all models is effected by tilting the headlamp shell after slackening the bolt that passes through the slotted adjuster bracket beneath the headlamp shell. The horizontal alignment of the beam can be adjusted by altering the position of the screw which passes through the headlamp rim. The screw is fitted in the 9 o'clock position when viewed from the front of the machine. Turning the screw in a clockwise direction will move the beam direction over to the left-hand side.

4   In the UK, regulations stipulate that the headlamp must be arranged so that the light will not dazzle a person standing in a distance greater than 25 feet from the lamp, whose eye level is not less than 3 feet 6 inches above that plane. It is easy to approximate this setting by placing the machine 25 feet away from a wall, on a level road, and setting the beam height so that it is concentrated at the same height as the distance of the centre of the headlamp from the ground. The rider must be seated normally during this operation and also the pillion passenger, if one is carried regularly.

16 Stop and tail lamp: replacement of bulbs

1   The combined stop and tail lamp bulb contains two filaments, one for the stop lamp and one for the tail lamp.

2   The offset pin bayonet fixing bulb can be renewed after the plastic lens cover and screws have been removed.

17 Flashing indicator lamps: replacing bulbs

1   Flashing indicator lamps are fitted to the front and rear of the machine. They are mounted on short stalks through which the wires pass. Access to each bulb is gained by removing the two screws holding the plastic lens cover. The bulbs are of 27W rating and are retained by a bayonet fixing.

14.2 Starter solenoid is mounted here

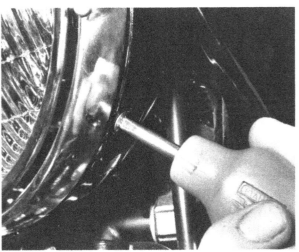

15.1a Release headlamp retaining screws ...

15.1b ... and lift unit away from shell

15.2a Connector pulls off sealed beam unit. Note pilot bulb

15.2b Rim halves are retained by self-tapping screws

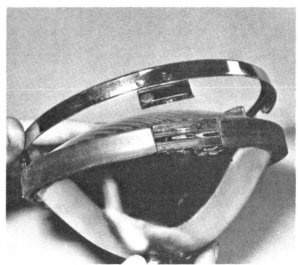

15.2c Separate halves to remove sealed beam unit

15.3a This screw controls up and down adjustment

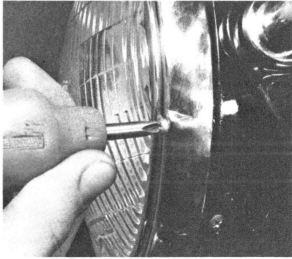

15.3b Screws in rim cater for lateral movement

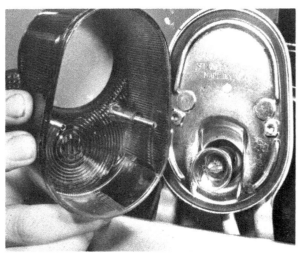

16.2 Remove plastic lens to gain access to bulb

17.1 Indicator bulb is a bayonet fit in holder

18 Flashing indicator relay: location and replacement

1   The flashing indicator relay, fitted in conjunction with the flashing indicator lamps is located beneath the frame top tube near the coils. It is mounted in a rubber 'box' which isolates it from the harmful effects of vibration.
2   When the relay malfunctions, it must be renewed; a repair is impracticable. When the unit is in working order audible clicks will be heard which coincide with the flash of the indicator lamps. If the lamps malfunction, check firstly that a bulb has not blown, or the handlebar switch is not faulty. The usual symptom of a fault is one initial flash before the unit goes dead.
3   Take great care when handling a flasher unit. It is easily damaged, if dropped.

19 Speedometer and tachometer heads: replacement of bulbs

1   Bulbs fitted to each instrument illuminate the dials during the hours of darkness when the headlamp is switched on. All bulbs fitted to either instrument head have the same type of bulb holder which is a push fit in the instrument base.
2   Access to the bulbs and holders is gained by removing the nuts and washers which secure the rubber mounted instruments to their individual pods. Lift the instruments clear of their pods and pull out the bulb holders.

20 Indicator panel lamps

1   An indicator lamp panel which holds three warning bulbs is fitted between the speedometer and tachometer heads. The panel is held to the instrument mounting bracket by screws from underneath. The bulbs are fitted in a holder which is a push fit in the panel base.

21 Horn: adjustment

1   The horn is mounted on a spring plate between the two frame down tubes. After considerable use the contacts inside the horn will wear. To compensate for wear an adjusting screw is

fitted at the rear of the horn. If the horn tone becomes inaudible or poor, turn the screw in slowly until the tone is correct again. Do not turn the screw in too far or the current increase may burn out the horn coil.

22 Ignition switch: removal and replacement

1   The combined ignition and lighting master switch is mounted in the warning light panel mounting plate.
2   If the switch proves defective it can be removed by releasing the warning light panel.
Remove the ignition switch nut and take off the upper switch cover. Disconnect the switch wiring socket. The switch and lower cover can be removed after unscrewing the two mounting bolts.
3   Reassembly of the switch can be made in the reverse procedure as described for dismantling. Repair is rarely practicable. It is preferable to purchase a new switch unit, which will probably necessitate the use of a different key.

23 Stop lamp switch: adjustment

1   All models have a stop lamp switch fitted to operate in conjunction with the rear brake pedal. The switch is located immediately to the rear of the crankcase, on the right-hand side of the machine. It has a threaded body giving a range of adjustment.
2   If the stop lamp is late in operating, slacken the locknuts and turn the body of the lamp in an anticlockwise direction so that the switch rises from the bracket to which it is attached. When the adjustment seems near correct, tighten the locknuts and test.
3   If the lamp operates too early, the locknuts should be slackened and the switch body turned clockwise so that it is lowered in relation to the mounting bracket.
4   As a guide, the light should operate after the brake pedal has been depressed by about 2 cm (¾ inch).
5   The front brake stop lamp switch is built into the hydraulic system and contains no provision for adjustment. If the switch malfunctions, it must be renewed.
6   Note that if the switch is removed from the hydraulic fluid distributor block, it will be necessary to bleed the hydraulic system on reassembly.

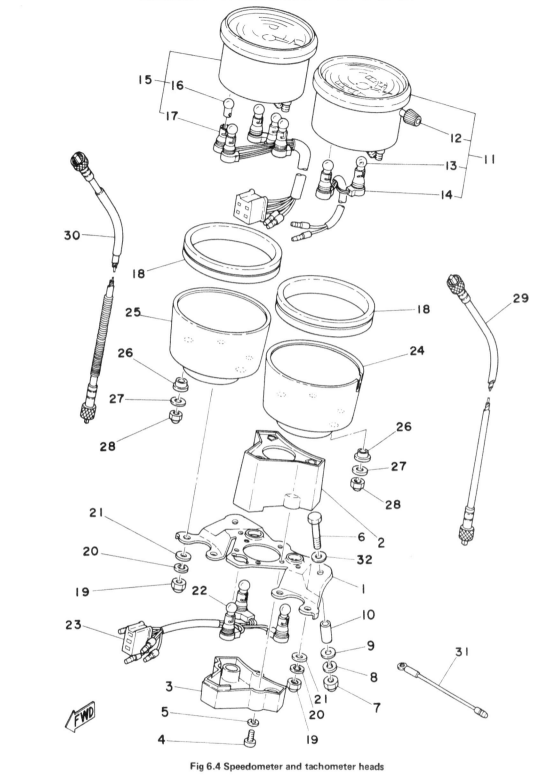

Fig 6.4 Speedometer and tachometer heads

1 Mounting bracket	12 Trip counter reset knob	23 Bulb holder and wiring assembly
2 Warning light housing	13 Bulb - 2 off	24 Speedometer housing
3 Warning light housing cover	14 Bulb holder	25 Tachometer housing
4 Screw - 3 off	15 Tachometer assembly	26 Rubber mounting - 4 off
5 Spring washer - 3 off	16 Bulb - 5 off	27 Plain washer - 4 off
6 Bolt - 2 off	17 Bulb holder assembly	28 Domed nut - 4 off
7 Domed nut - 2 off	18 Damping ring - 2 off	29 Speedometer cable
8 Spring washer - 2 off	19 Domed nut - 4 off	30 Tachometer cable
9 Plain washer - 2 off	20 Spring washer - 4 off	31 Earth lead
10 Spacer - 2 off	21 Plain washer - 4 off	32 Plain washer - 2 off
11 Speedometer assembly	22 Bulb - 3 off	

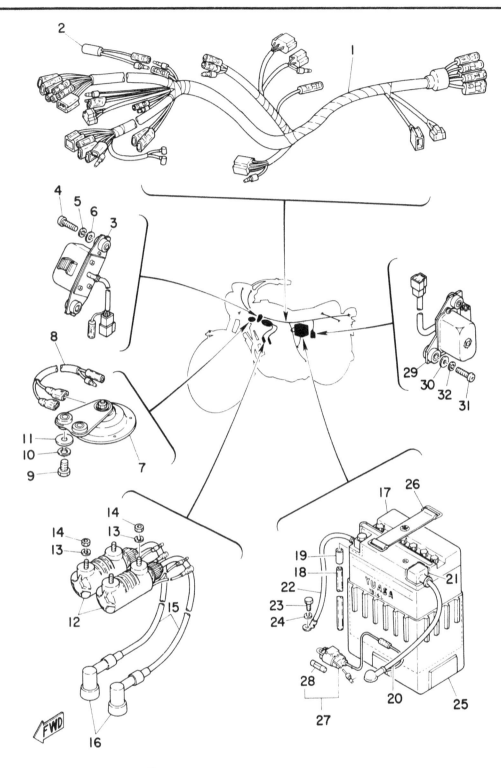

Fig. 6.5. Electrical system - component location

1	Wiring harness	12	Ignition coil - 2 off	23	Bolt
2	Diode	13	Spring washer - 4 off	24	Spring washer
3	Relay	14	Nut - 4 off	25	Battery seat
4	Screw - 2 off	15	HT lead - 2 off	26	Retaining strap
5	Spring washer - 2 off	16	Plug cap - 2 off	27	Fuse holder
6	Plain washer - 2 off	17	Battery	28	Fuse
7	Horn	18	Drain hose	29	Voltage regulator
8	Horn leads	19	Collar	30	Plain washer - 2 off
9	Bolt - 2 off	20	Battery + lead	31	Screw - 2 off
10	Spring washer - 2 off	21	Rubber cover	32	Spring washer - 2 off
11	Plain washer - 2 off	22	Earth lead		

20.1 Note location of warning lamps and ignition switch

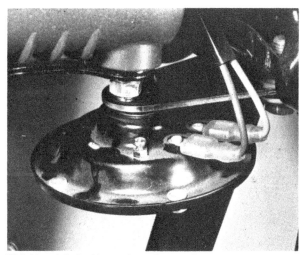

21.1 Horn is bolted beneath top frame tube

23.1a Rear brake switch is adjustable

23.1b This switch indicates maximum brake wear

24 Handlebar switches: general

1   Generally speaking, the switches give little trouble, but if necessary they can be dismantled by separating the halves which form a split clamp around the handlebars. Note that the machine cannot be started until the ignition cut-out on the right-hand end of the handlebars is turned to the central 'ON' position.
2   Always disconnect the battery before removing any of the switches, to prevent the possibility of a short circuit. Most troubles are caused by dirty contacts, but in the event of the breakage of some internal part, it will be necessary to renew the complete switch.
3   The best method of maintaining the switches is to use one of the proprietary aerosol switch cleaning fluids, which will restore the condition of the contacts and will help prevent subsequent corrosion.

24.1a Main electrical functions are controlled by ...

24.1b ... left and right-hand switch units

25 Fault diagnosis: electrical system

Symptom	Cause	Remedy
Complete electrical failure	Blown fuse	Check wiring and electrical connections for short circuit before fitting new fuse.
	Isolated battery	Check battery connections for break or corrosion. Clean, smear with petroleum jelly and retighten connections.
	Serious short to earth	Check wiring for breaks in insulation and renew or repair as necessary. Check connections for signs of water. Use water dispersal aerosol to remedy.
Dim lights, horn and starter inoperative	Discharged battery	Check battery with hydrometer, and recharge as necessary. If problem occurs frequently, check output of alternator, and peripheral components. Check battery for weak cell or sulphation.
Constantly blowing bulbs	Vibration, poor earth connections	Check and clean bulb holders and cables. Check earth return connections. Tighten loose fittings.
Starter motor sluggish	Worn brushes	Remove motor and check brushes. Clean or renew as necessary.
	Discharged battery	See above
	Faulty battery	Check for faulty cell or sulphation. Check that terminals are not corroded.
Starter motor inoperative	Faulty solenoid	Check battery condition. If fault persists, renew solenoid.
Starter motor fails spasmodically	'Dead' commutator segment	Remove and dismantle starter motor. Check segments with multimeter.
Light dim or flickering	Resistence in circuit	Check all connections. Check and clean switch contacts and bulb holders.
Indicators flash once then stop	Indicator bulb blown or badly connected	Renew if necessary, clean connections and check earth

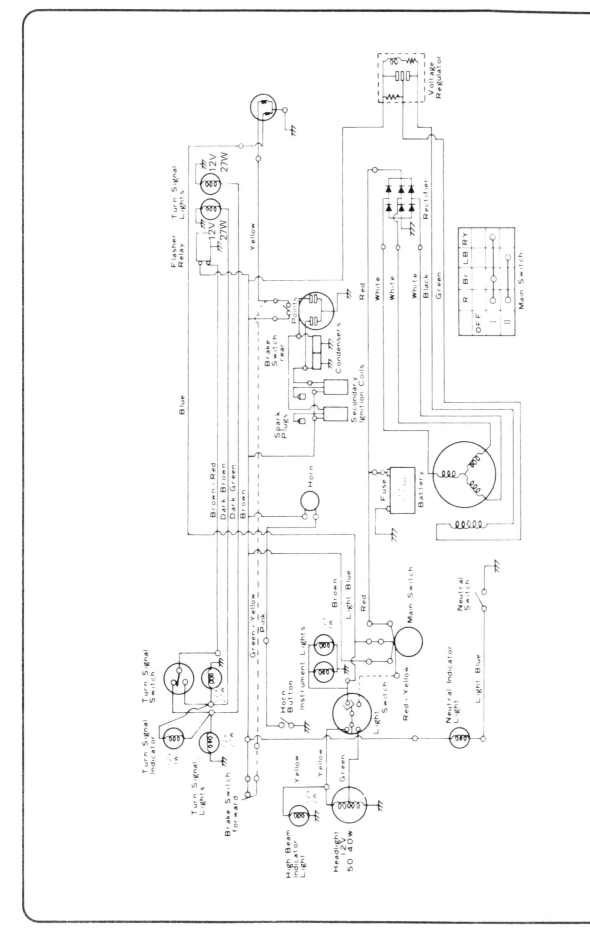

Wiring diagram XS1B

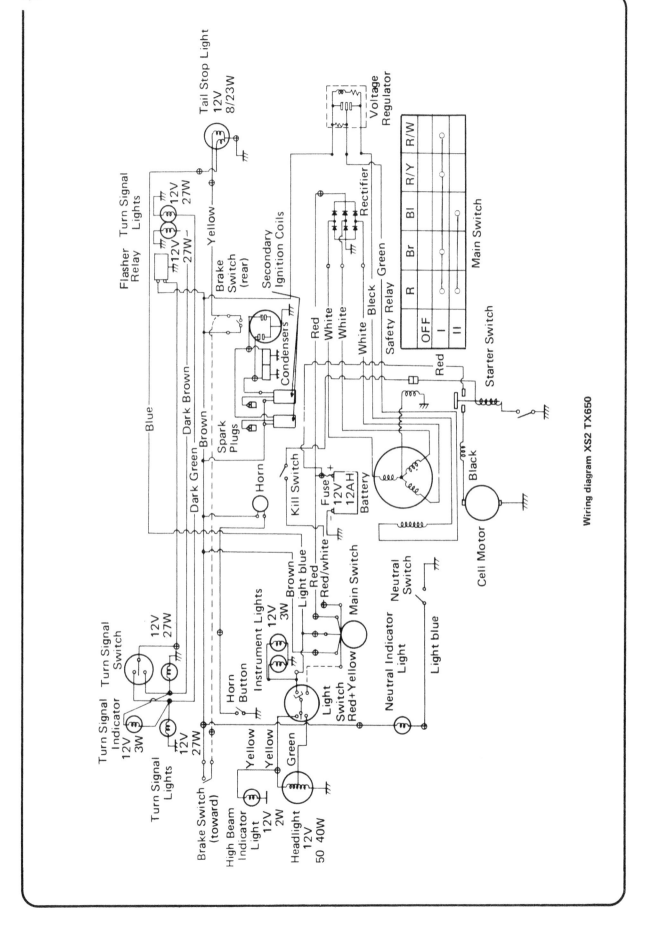

Wiring diagram XS2 TX650

135

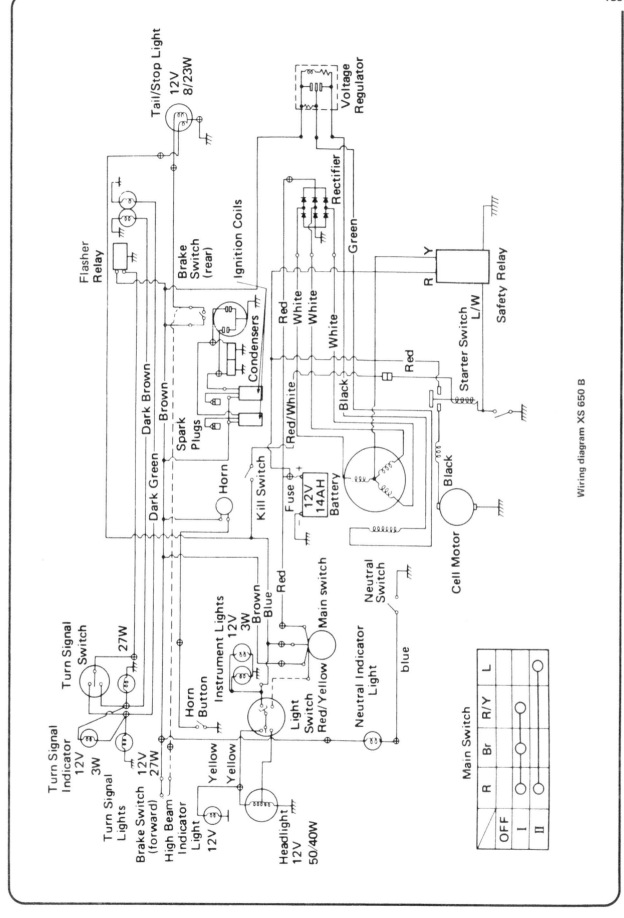

Wiring diagram XS 650 B

1980 Yamaha XS650 Special model

Engine unit of the XS650 Special model

Chapter 7 The 1977 to 1983 models

Contents

Specifications

Read for all models unless otherwise stated

Engine, clutch and gearbox specifications

Engine

Bore	75 mm (2.953 in)
Stroke	74 mm (2.913 in)
Capacity	653 cc (39.8 cu in)
Compression ratio:	
XS650 (UK) and XS650 D (US)	8.4 : 1
XS650 SE (UK), XS650 E, F, SE, SF, 2F (US)	8.5 : 1
XS650 G, H, SG, SH, SJ, SK (US)...	8.7 : 1

Crankshaft

End deflection limit	0.05 mm (0.002 in)
Con-rod big-end to flywheel clearance	0.15 - 0.4 mm (0.0059 - 0.0157 in)

Cylinder barrel

Standard bore size	75.00 - 75.02 mm (2.9528 - 2.9536 in)
Wear limit	75.10 mm (2.957 in)
Taper limit..	0.05 mm (0.002 in)
Out of round limit	0.01 mm (0.0004 in)

Cylinder head

Gasket thickness..	1.20 mm (0.047 in)
Warp limit	0.03 mm (0.002 in)

Pistons

Skirt clearance	0.050 - 0.055 mm (0.0020 - 0.0022 in)
Oversizes...	75.25 mm (2.963 in), 75.50 mm (2.972 in), 75.75 mm (2.982 in), 76.00 mm (2.992 in)
Gudgeon pin:	
Outer diameter	19.995 - 20.000 mm (0.7898 - 0.7900 in)
Length..	60.70 - 61.0 mm (2.3884 - 2.40 in)

Piston rings

End gap (installed):	
Top and second	0.2 - 0.4 mm (0.008 - 0.016 in)
Oil ring..	0.3 - 0.9 mm (0.012 - 0.035 in)
Side clearance:	
Top.......	0.04 - 0.08 mm (0.0016 - 0.0031 in)
Second.	0.03 - 0.07 mm (0.0012 - 0.0028 in)

Camshaft

Cam lift:	
Inlet	7.991 mm (0.315 in)
Exhaust	8.030 mm (3.161 in)
Cam height:	
Inlet	39.99 $\pm$ 0.05 mm (1.574 $\pm$ 0.002 in)
Exhaust	40.03 $\pm$ 0.05 mm (1.576 $\pm$ 0.002 in)
Cam width:	
Inlet	32.24 $\pm$ 0.05 mm (1.269 $\pm$ 0.002 in)
Exhaust	32.30 $\pm$ 0.05 mm (1.272 $\pm$ 0.002 in)
Run-out limit	0.03 mm (0.002 in)
Chain:	
Type.....	Tsubakimoto BFO5M
Pitch.....	7.774 mm (0.303 in)
No of links	106 (including joint)
Camshaft sprocket:	
Ratio.....	2.000 : 1
No of teeth	36/18

Rocker arms and shafts

Rocker arm inner diameter	15.000 - 15.018 mm (0.5910 - 0.5917 in)
Rocker shaft outer diameter	14.985 - 14.991 mm (0.5904 - 0.5907 in)
Clearance	0.009 - 0.033 mm (0.00035 - 0.0013 in)

Valves

Stem diameter - inlet:

All UK models and XS650 E, F, SE, SF, 2F (US) .. .. 7.975 - 7.990 mm (0.3141 - 0.3146 in)
XS650 G, H, SG, SH, SJ, SK (US)... .... 7.985 - 8.000 mm (0.3140 - 0.3150 in)

Stem diameter - exhaust .... 7.960 - 7.975 mm (0.3134 - 0.3141 in)

Guide bore diameter:

Inlet and exhaust .... 8.010 - 8.019 mm (0.3154 - 0.3157 in)

Stem to guide bore clearance - inlet:

All UK models and XS650 E, F, SE, SF, 2F (US) .. .. 0.020 - 0.044 mm (0.00079 - 0.00173 in)
XS650 G, H, SG, SH, SJ, SK (US)... .... 0.010 - 0.034 mm (0.0004 - 0.0013 in)

Stem to guide bore clearance - exhaust .... 0.035 - 0.059 mm (0.00138 - 0.00232 in)

Head diameter:

Inlet .... 41 mm (1.614 in)
Exhaust .... 35 mm (1.378 in)

Seat width - inlet and exhaust... .... 1.30 mm (0.051 in)

Face width - inlet and exhaust .. .. 2.10 mm (0.083 in)

Margin thickness - inlet and exhaust ... . 1.30 mm (0.051 in)

Valve clearances (engine cold)

Inlet:

XS650 (UK), XS650 D (US) ... . 0.05 mm (0.002 in)
XS650 E and SE (US) .... 0.10 mm (0.0039 in)
XS650 SE (UK), XS650 F, G, H, SF, 2F, SG, SH, SJ, SK (US) ... 0.06 mm (0.0024 in)

Exhaust .... 0.15 mm (0.0059 in)

Valve timing

Inlet - all UK models and XS650 D, E, F, SE, SF, 2F (US):

Opens .. .. 36° BTDC
Closes.. .... 68° ABDC
Duration .... 284°
Overlap .... 72°

Inlet - XS650 G, H, SG, SH, SJ, SK (US):

Opens .. .. 35° BTDC
Closes.. .... 69° ABDC
Duration .... 284°
Overlap .... 72°

Exhaust - all UK models and XS650 D, E, F, SE, SF, 2F (US):

Opens .. .. 68° BBDC
Closes.. .... 36° ATDC
Duration .... 284°
Overlap .... 72°

Exhaust - XS650 G, H, SG, SH, SJ, SK (US):

Opens .. .. 67° BBDC
Closes.. .... 37° ATDC
Duration .... 284°
Overlap .... 72°

Valve springs

Free length (inlet and exhaust):

Inner..... .... 42.0 mm (1.654 in)
Outer.... .... 42.5 mm (1.675 in)

Clutch

Friction plate:

No off ... . 7 early models, 6 later models
Thickness .... 3 mm (0.118 in)
Wear limit .... 2.7 mm (0.106 in)
Warp limit .... 0.05 mm (0.002 in)

Clutch plate:

No off ... . 6 early models, 5 later models
Thickness .... 1.4 mm (0.055 in)
Warp limit .... 0.05 mm (0.002 in)

Spring:

No off 6
Free length .... 34.6 mm (1.362 in)
Wear limit .... 33.6 mm (1.323 in)

Clutch housing radial play .... 0.027 - 0.081 mm (0.0011 - 0.0032 in)

Pushrod bending limit..... .... 0.2 mm (0.008 in)

Gearbox

Gear ratios (No of teeth):

1st	2.461 : 1 (32/13T)
2nd	1.588 : 1 (27/17T)
3rd	1.300 : 1 (26/20T)
4th	1.095 : 1 (23/21T)
5th	0.956 : 1 (22/23T)

Torque settings - lbf ft (kgf m)	XS650 (UK)	XS650 D (US)	XS650 E (US)	XS650 SE (UK), XS650 F, G, H, SE, SF, 2F, SG, SH, SJ, SK (US)
Cylinder head:				
10 mm nut	24 (3.3)	22 - 25 (3.0 - 3.5)	27 (3.7)	27 (3.7)
8 mm bolt	16.6 (2.3)	15 - 18 (2.1 - 2.5)	16 (2.2)	15 (2.1)
6 mm bolt	9.4 (1.3)	7.2 - 10.8 (1.0 - 1.5)	7 (1.0)	6.5 (0.9)
10 mm stud bolt	13 (1.8)	11 - 14.5 (1.5 - 2.0)	-	-
Cylinder head side cover:				
6 mm crown nut	-	-	7.0 (1.0)	6.5 (0.9)
8 mm crown nut	-	-	10 (1.5)	9.5 (1.3)
Spark plug	14.5 (2.0)	11 - 18 (1.5 - 2.5)	14.5 (2.0)	14.5 (2.0)
Alternator nut	50 (7.0)	36 - 58 (5.0 - 8.0)	20 (4.0)	27.5 (3.8)
ATU bolt (6 mm)	-	-	6 (0.8)	6 (0.8)
Cam chain tensioner cap	-	-	16 (2.2)	15 (2.1)
Oil drain plug	27.5 (3.8)	25 - 29 (3.5 - 4.0)	32 (4.4)	30.5 (4.2)
Oil filter bolt	-	-	7.0 (1.0)	6.5 (0.9)
Oil delivery pipe union bolt:				
10 mm	15 (2.1)	14.5 - 16 (2.0 - 2.2)	16 (2.2)	15 (2.1)
14 mm	20 (2.8)	18 - 22 (2.5 - 3.0)	-	-
Exhaust pipe retaining nut	-	-	11 (1.5)	9.5 (1.3)
Crankcase bolt/nut (8 mm)	14.5 (2.0)	14.5 (2.0)	16 (2.2)	15 (2.1)
Engine mountings:				
Upper nut (8 mm)	13 (1.8)	10 - 16 (1.4 - 2.2)	14 (2.0)	13 (1.8)
Upper nut (10 mm)	21.5 (3.0)	17 - 27 (2.3 - 3.7)	36 (5.0)	21.5 (3.0)
Front nut (10 mm)	33.5 (4.6)	25 - 40 (3.5 - 5.6)	36 (5.0)	33.5 (4.6)
Rear nut (10 mm)	29 (4.0)	22 - 36 (3.0 - 5.0)	33 (4.5)	29.5 (4.1)
Rear lower nut (10 mm)	33.5 (4.6)	25 - 40 (3.5 - 5.6)	36 (5.0)	33.5 (4.6)
Lower nut (10 mm)	33.5 (4.6)	25 - 40 (3.5 - 5.6)	36 (5.0)	65 (9.0)
Clutch centre nut	50 (7.0)	36 - 58 (5.0 - 8.0)	47 (6.5)	58 (8.0)
Primary drive gear nut	65 (9.0)	58 - 82 (8.0 - 10.0)	65 (9.0)	65 (9.0)
Final drive sprocket nut	79 (11.0)	72 - 87 (10.0 - 12.0)	36 (5.0)	47 (6.5)

Fuel and lubrication specifications

Fuel tank capacity

XS650 (UK), XS650 D, E, F (US)	15.0 lit (3.30 Imp gal, 3.96 US gal)
XS650 SE (UK), early XS650 SE, XS650 G, SG, H, SH, SJ, SK (US)	11.5 lit (2.53 Imp gal, 3.04 US gal)
XS650 SE post engine No 2F0 -114 241, SF, 2F (US)	11.0 lit (2.42 Imp gal, 2.90 US gal)

Fuel grade

Unleaded or low-lead regular

Carburettors

	XS650 (UK)	XS650 D (US)
Make/type	Mikuni BS38	Mikuni BS38
ID mark	-	58400
Main jet	117.5	122.5
Jet needle - clip position	4M1-3	4M1-3
Needle jet	Z-8	Z-8
Pilot jet	25	25
Pilot screw (turns out)	1 1/4	1 1/4
Idle speed	1200 rpm	1200 rpm
Vacuum synchronisation	Same readings	Same readings
Float height	25 ± 2.5 mm (0.98 ± 0.098 in)	25 ± 2.5 mm (0.98 ± 0.098 in)

Carburettors

	XS650 SE (UK)	XS650 E, F, SE, SF, 2F (US)
Make/type	Mikuni BS34	Mikuni BS38
ID mark	3L100	2F000

Carburettors (continued)

	XS650 SE (UK)	XS650 E, F, SE, SF, 2F (US)
Main jet	132.5	135
Jet needle - clip position	5HX13-3	502-3
Needle jet	Y-0	Z-2
Air jet	85	140
Pilot jet	42.5	27.5
Pilot screw (turns out)	Preset	Preset
Idle speed..	1200 rpm	1200 rpm
Vacuum synchronisation.	Same readings	Same readings
Float height	24 ± 1 mm (0.94 ± 0.04 in)	24 ± 1 mm (0.94 ± 0.04 in)

Carburettors

	XS650 G, H, SG, SH, SJ, SK (US)
Make/type	Mikuni BS34
ID mark	3G100, 5V400 on SJ, SK
Main jet	132.5
Jet needle..	5HX12
Needle jet	Y-0
Air jet	85
Pilot jet	42.5
Pilot screw (turns out)	Preset
Idle speed..	1200 rpm
Vacuum synchronisation.	Same readings
Float height:	
XS650 G, SG models	27.3 ± 0.5 mm (1.075 ± 0.02 in)
XS650 H, SH, SJ, SK models	22.0 ± 1.0 mm (0.866 ± 0.04 in)
Fuel level - H, SH, SJ, SK models	1 ± 1 mm (0.04 ± 0.04 in)

Air cleaners

Type	Dry, foam rubber
Element cleaning interval:	
XS650 E and SE (US)	1000 miles (1600 km)
XS650 SE (UK)	2000 miles (3000 km)
XS650 SE (post engine No 2F0 - 114 241), SF, 2F, G, H, SG, SH, SJ, SK (US)	5000 miles (8000 km)

Lubrication system

Engine oil capacity:	
Oil quantity at rebuild	2.5 lit (4.4 Imp pt, 5.28 US pt)
Oil change quantity	2.0 lit (3.5 Imp pt, 4.23 US pt)
Oil pump:	
Top clearance	0.10 - 0.18 mm (0.0039 - 0.0071 in)
Tip clearance	0.03 - 0.09 mm (0.0012 - 0.0035 in)
Side clearance	0.03 - 0.08 mm (0.0012 - 0.0031 in)
Bypass valve setting pressure	14 psi (1.0 kg/cm^2)

Ignition system specifications

Ignition timing .

....	15° BTDC at 1200 rpm

Spark plug

Type:	
XS650 (UK)	NGK BP8ES
XS650 SE (UK) and all US models..	NGK BP7ES or Champion N7Y
Gap	0.7 - 0.8 mm (0.027 - 0.031 in)

Ignition coil

Spark gap:	
US models	6 mm (0.24 in) or more at 500 rpm
UK models	8 mm (0.31 in) or more at 300 rpm at 8 volts
Primary winding resistance:	
All UK models, XS650 E, F, SE, SF, 2F (US)	3.9 ohms ± 10% at 20°C (68°F)
XS650 G, H, SG, SH, SJ, SK (US)...	2.5 ohms ± 10% at 20°C (68°F)
Secondary winding resistance:	
All UK models, XS650 E, F, SE, SF, 2F (US)	8 K ohms ± 20% at 20°C (68°F)
XS650 G, H, SG, SH, SJ, SK (US)...	13 K ohms ± 20% at 20°C (68°F)

Ignition system specifications (continued)

Contact breakers - all UK models and US XS650 D, E, F, SE, SF, 2F

Points gap	0.30 - 0.40 mm (0.012 - 0.016 in)
Point spring pressure	650 - 850 grammes (22.9 - 30.0 oz)
Cam closing angle	93° ± 5°

Condensers - all UK models and US XS650 D, E, F, SE, SF, 2F

Capacity	0.22 mF
Insulation resistance	10 M ohms or more

Frame and forks specifications

Front forks

Spring free length:
XS650 (UK), XS650 D (US)	470.5 mm (18.5 in)
XS650 SE (UK), XS650 E, F, G, H, SE, SF, 2F, SG, SH, SJ, SK (US)	482 ± 6 mm (18.98 ± 0.24 in)
Fork travel	150 mm (5.906 in)
Stanchion outer diameter	35 mm (1.378 in)
Damping oil capacity	169 cc (5.9 Imp fl oz, 5.7 US fl oz)

Damping oil grade:
UK models	10W/30 SE motor oil
US models	10W fork oil

Rear suspension

Spring free length.	226 mm (8.90 in)
Swinging arm free play (max)	1 mm (0.04 in)

Final drive chain

Size:
XS650 (UK), XS650 D (US)	DK530 HDS/Daido
XS650 SE (UK), XS650 E, F, G, H, SE, SF, 2F, SG, SH, SJ, SK (US)	50 HDS

No of links:
XS650 SE (UK)	105 links + joint
All other models	103 links + joint
Pitch	15.875 mm (5/8 in)
Free play....	20 - 30 mm (0.78 - 1.17 in)

Torque settings - lbf ft (kgf m)	XS650 (UK)	XS650 D (US)	XS650 E (US)	XS650 SE (UK), XS650 F, G, H, SE, SF, 2F, SG, SH, SJ, SK (US)
Front wheel spindle nut	65 (9.0)	51 - 72 (7.0 - 10.0)	61 (8.5)	77.5 (10.7)
Front wheel spindle clamp	10 (1.4)	7 - 12 (1.0 - 1.7)	11 (1.5)	10 (1.4)
Upper yoke fork clamp bolts	7.2 (1.0)	6 - 9 (0.8 - 1.2)	7 (1.0)	8 (1.1)
Lower yoke fork clamp bolts	7.2 (1.0)	6 - 9 (0.8 - 1.2)	11 (1.5)	14.5 (2.0)
Steering stem crown bolt	40 (5.5)	30 - 47 (4.2 - 6.5)	40 (5.5)	39 (5.4)
Upper yoke pinch bolt	7 (1.0)	6 - 9 (0.8 - 1.2)	7 (1.0)	7 (1.0)
Rear suspension unit bolts:				
Upper...	21.5 (3.0)	17 - 27 (2.3 - 3.7)	-	21.5 (3.0)
Lower...	29 (4.0)	22 - 35 (3.0 - 4.8)	22 (3.0)	28 (3.9)
Swinging arm pivot shaft nut.....	50 (7.0)	36 - 58 (5.0 - 8.0)	47 (6.5)	47 (6.5)
Rear wheel spindle nut	108 (15.0)	87 - 130 (12.0 - 18.0)	108 (15.0)	108 (15.0)

Wheels, brakes and tyres specifications

Wheels

Rim size:	Front	Rear
XS650 (UK), XS650 D, E, F (US)	1.85 x 19	2.15 x 18
XS650 SE (UK), XS650 SE, SF, SG, SH (US)	1.85 x 19	3.00 x 16
XS650 G, H, 2F, SJ, SK (US)	1.85 x 19	2.75 x 16
Rim runout - radial and axial	2.0 mm (0.08 in)	2.0 mm (0.08 in)

Tyres

	Front	Rear
Size:		
XS650 (UK), XS650 D, E, F (US)	3.50H19 - 4PR	4.00H18 - 4PR
XS650 SE (UK), XS650 G, H, SE, SF, 2F,		
SG, SH (US)*...	3.50S19 - 4PR	130 - 90S16 - 4PR
XS650 SJ, SK (US)	3.50S19 - 4PR	130/9067S

XS650 SF, SG and SH models are suitable for tubeless type tyres

Tyre pressures - psi (kg/cm²)

Up to 90 kg (198 lb) load	22 (1.6)	28 (2.0)
90 - 206 kg (198 - 453 lb) max load*	28 (2.0)	32 (2.3)

Use these pressures for high speed riding

Disc brakes

Disc thickness	7.0 mm (0.276 in)
Wear limit	6.5 mm (0.256 in)
Disc pad thickness	11.0 mm (0.430 in)
Wear limit	6.0 mm (0.240 in)
Brake fluid type....	DOT 3 or 4
Master cylinder bore inner diameter	14.0 mm (0.551 in)
Caliper bore inner diameter	38.1 mm (1.500 in)

Drum brake

Brake drum diameter	180 mm (7.087 in)
Lining thickness	4 mm (0.160 in)
Wear limit	2 mm (0.080 in)
Spring free length.	68 mm (2.68 in)

Torque settings - lbf ft (kgf m)

	XS650 (UK)	XS650 D (US)	XS650 E (US)	XS650 SE (UK), XS650 F, G, H, SE, SF, 2F, SG, SH, SJ, SK (US)
Front wheel spindle nut	65 (9.0)	51 - 72 (7.0 - 10.0)	61 (8.5)	77.5 (10.7)
Front wheel spindle clamp	10 (1.4)	7 - 12 (1.0 - 1.7)	11 (1.5)	10 (1.4)
Brake disc to hub bolt	14 (2.0)	12 - 16 (1.7 - 2.2)	14 (2.0)	14 (2.0)
Caliper to bracket bolt	13 (1.8)	11 - 14.5 (1.5 - 2.0)	14 (2.0)	13 (1.8)
Master cylinder hose union bolt	18.8 (2.6)	17 - 20 (2.3 - 2.8)	18 (2.5)	18.8 (2.6)
Rear wheel spindle nut	108 (15.0)	87 - 130 (12.0 - 18.0)	108 (15.0)	108 (15.0)

Electrical specifications

Battery

Type	YUASA YB14L - A2
Capacity	12V, 14Ah
Charging rate	1.4 amps/10 hours
Specific gravity.....	1.28 at 20°C (68°F)

Alternator

Charging output:	
All UK models, XS650 D, E, F, SE, SF, 2F (US)....	14 volts, 11 amperes at 2000 rpm
XS650 G, H, SG, SH, SJ, SK (US)...	14 volts, 16 amperes at 5000 rpm
Rotor coil resistance (field coil)..	5.25 ± 10% ohms at 20°C (68°F)
Stator coil resistance	0.46 ± 10% ohms at 20°C (68°F)
Brush length	14.5 mm (0.571 in)
Brush wear limit	7.0 mm (0.276 in)

Starter motor

Output	0.5 kw
Armature coil resistance	0.0067 ohms ± 10% at 20°C (68°F)
Field coil resistance	0.004 ohms ± 10% at 20°C (68°F)
Brush:	
Length..	16 mm (0.63 in)
Wear limit	4 mm (0.16 in)
Spring pressure	800 grammes (28.2 oz)
Commutator:	
Diameter	33 mm (1.30 in)
Wear limit	30 mm (1.18 in)
Mica undercut.	0.7 mm (0.028 in)

Electrical specifications (continued)

Starter switch

Amperage rating	100 amps
Winding resistance	3.5 ohms
Starter clip friction tension	2.2 - 2.5 kg (4.9 - 5.5 lb)

Regulator - all UK models, XS650 D, E, F, SE, SF, 2F (US)

Type	Hitachi TLIZ - 80 (Tillil type)
Regulating voltage	14.5 $\pm$ 0.5 volts
Core gap	0.6 - 1.0 mm (0.024 - 0.039 in)
Point gap...	0.3 - 0.4 mm (0.012 - 0.016 in)
Voltage coil resistance....	10 ohms at 20°C (68°F)
Resistor value	10 - 25 ohms at 20°C (68°F)

Rectifier - all UK models, XS650 D, E, F, SE, SF, 2F (US)

Type	Hitachi SB6B - 17 (6-element type, full-wave)
Capacity	12 amps

Regulator/rectifier - XS650 G, H, SG, SH, SJ, SK (US)

Type	Toshiba S8515 (IC type)
Regulated voltage	14.5 $\pm$ 0.3 volts

Fuse ratings

Main fuse...	20 A
Circuit fuses - XS650 G, H, SE, SF, 2F, SG, SH, SJ, SK:	
Headlamp	10 A
Ignition	10 A
Signal	10 A

Bulbs

Headlamp..	12V 50/40W (sealed beam unit)
Flashing indicator lamps.	12V 27W (21W later UK models)
Licence plate light - later models	12V 3.8W
Tail/stop lamp:	
US models	12V 8/27W
XS650 (UK)	12V 7/23W
XS650 SE (UK)	12V 5/21W
Speedometer/tachometer lamps	12V 3.4W
Pilot lamps (turn signal, main beam, neutral, headlamp failure, tail/stop lamp failure - as applicable)	12V 3.4W

---

1 Introduction

UK models

There have been relatively few versions of the XS650 in the UK. The standard XS650 covered in Chapters 1 to 6 of this book continued with only minor modification until discontinued in June 1981. The custom-styled XS650 SE (US Custom/Special) was introduced in March 1979. Changes to the SE were few, with a revised tail lamp assembly on the 1980 model and the reversion to drum braking at the rear on the 1981 model. The SE was discontinued in April 1982.

US models

This Chapter covers US models from the 1977 model year, until their discontinuation at the end of 1983. Model suffix letters are used to denote the production year, and these are given in the table below.

    1977   XS650 D (standard)
    1978   XS650 E (standard) and SE (Special)
    1979   XS650 F (standard), SF (Special) and 2F (Special II)
    1980   XS650 G (Special II) and SG (Special)
    1981   XS650 H (Special II) and SH (Special)
    1982   XS650 SJ (Heritage Special)
    1983   XS650 SK (Heritage Special)
Standard specification models include the XS650 D, E and F,

these being a continuation of the XS650 C covered in the previous chapters; they were discontinued at the end of 1979. The 'Special' models introduced in 1978 had custom-styling, cast alloy wheels and a rear disc brake. The Special II differed from the Special model in having spoked wheels and a drum rear brake; both models were fitted with transistor controlled ignition in 1980. The Special reverted to having a rear drum brake in 1981 and for 1982 it underwent a change of name to 'Heritage Special' and was fitted with wire-spoked wheels and featured revised footrest mountings.

How to use this Supplement

When working on one of the models described above, refer first to this Chapter. If the information required is not given here, it can be assumed that the task is the same as for the earlier models in Chapters 1 to 6. Models are identified by their model suffix letters and, where necessary, by their production year.

---

2 Routine Maintenance

Topping up the engine oil - later UK models and XS650 H, SH, SJ and SK US models

1 These models are fitted with a sight glass in the right-hand

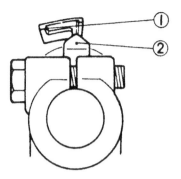

Fig. 7.1 Drum rear brake wear limit line (1) and indicator pointer (2) (Sec 2)

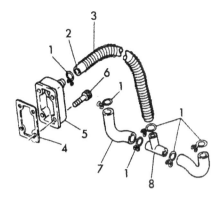

Fig. 7.2 Cylinder head breather assembly (Sec 3)

1	Clip	5	Breather cover
2	Hose	6	Allen bolt
3	Hose guard	7	Pipe
4	Gasket	8	Joint

crankcase cover. This is useful for keeping a regular watch on the oil level, but checking via the dipstick is still advised at scheduled intervals.

2   Place the motorcycle on its centre stand on level ground. Run the engine for a few minutes then stop it. Withdraw the dipstick, wipe its end and place it back into the crankcase (not screwed home). Withdraw the dipstick and check the level. If the oil level is correct it should lie within the two lines on the dipstick and similarly between the upper and lower lines on the sight glass.

3   Add oil of the recommended type via the dipstick hole if topping up is required.

Disc brake pad wear check and renewal

4   Pad wear can be checked via the inspection cap in the rear (front brake) or top (rear brake) of the caliper body. Pad renewal is required if the friction material has worn down level with the base of the central groove or the recessed edge of the pad material (depending on pad type); on some pads a red wear limit line will be visible. If there is any doubt about the amount of friction material remaining, remove the pads for direct inspection (see Sec 18). Always renew both pads at the same time, and if twin disc brakes are fitted renew those in both calipers.

Rear drum brake lining check - 1981-on UK models and XS650 H, SH, SJ and SK US models

5   If the motorcycle is fitted with a wear indicator pointer on the brake operating camshaft, the shoe lining condition can be assessed without removing the wheel for direct measurement. The pointer corresponds with a cast scale on the brake backplate. When the pointer approaches the wear limit line the

shoes should be renewed (see Fig. 7.1). Note that there is no real substitute for checking the shoes directly, especially if there is some doubt about brake performance.

---

3   Engine: modifications

Cylinder head breather assembly - later models

1   It should be noted when carrying out maintenance procedures on the cylinder head breather assembly, that on later models the design of the system has been changed. Reference should be made to Fig. 7.2.

Cam chain tension adjustment - later models

2   Although the procedure for adjusting the cam chain tension remains the same as that stated in Routine Maintenance at the front of this manual, it should be noted that on later models a gasket and nut are fitted under the adjuster cover. This nut is designed to act as a locknut for the adjuster bolt.

3   When fitting the adjuster cover note the torque settings given in the Specifications at the beginning of this Chapter.

Cylinder head tightening sequence - XS650 SJ and SK models

4   Note the revised nut/bolt tightening sequence shown in Fig. 7.4.

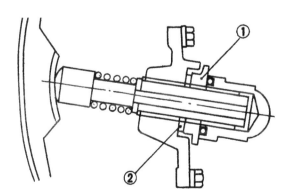

Fig. 7.3 Modified cam chain tensioner (Sec 3)

1	Locknut	2	Gasket

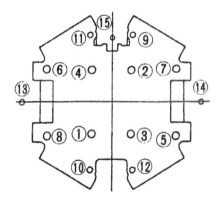

Fig. 7.4 Cylinder head tightening sequence - XS650 SJ and SK US models (Sec 3)

Engine earth lead - XS650 H, SH, SJ and SK US models

5   Note that the earth lead attached to one of the two cylinder head steady bracket bolts on earlier models is deleted on later US models.

Output shaft oil seal renewal

6   A modified oil seal was introduced during 1980 which enabled seal renewal without separating the crankcases. When purchasing a new seal, inspect its periphery first before attempting installation by this method (refer to Fig. 7.5 for identification details).

7   To fit the new type of seal, remove the final drive sprocket, lock washer and spacer from the output shaft, and pry out the old seal whilst taking care not to damage the seal housing or surrounding casing. Lubricate the new seal's outer face and drive into the casing using a suitable drift which bears only on its outer edges; stop when the seal face is flush with the casing - it must not be driven in any further.

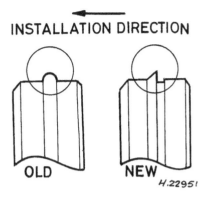

Fig. 7.5 Output shaft oil seal identification (Sec 3)

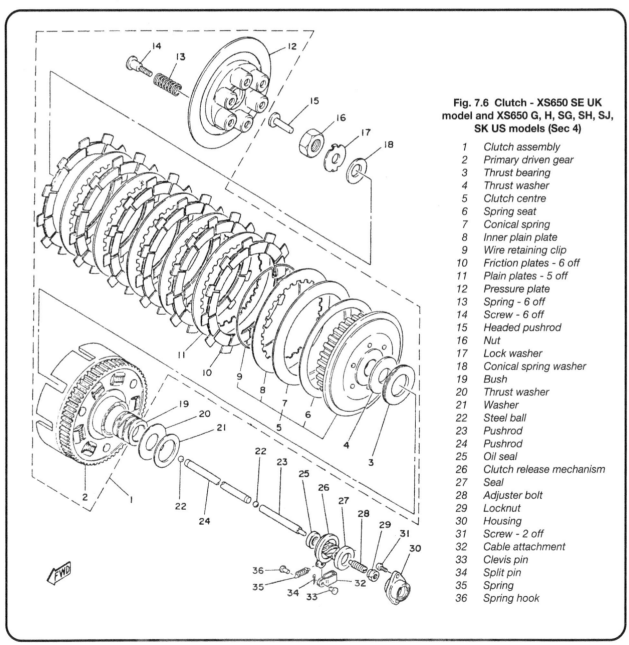

Fig. 7.6 Clutch - XS650 SE UK model and XS650 G, H, SG, SH, SJ, SK US models (Sec 4)

1	Clutch assembly
2	Primary driven gear
3	Thrust bearing
4	Thrust washer
5	Clutch centre
6	Spring seat
7	Conical spring
8	Inner plain plate
9	Wire retaining clip
10	Friction plates - 6 off
11	Plain plates - 5 off
12	Pressure plate
13	Spring - 6 off
14	Screw - 6 off
15	Headed pushrod
16	Nut
17	Lock washer
18	Conical spring washer
19	Bush
20	Thrust washer
21	Washer
22	Steel ball
23	Pushrod
24	Pushrod
25	Oil seal
26	Clutch release mechanism
27	Seal
28	Adjuster bolt
29	Locknut
30	Housing
31	Screw - 2 off
32	Cable attachment
33	Clevis pin
34	Split pin
35	Spring
36	Spring hook

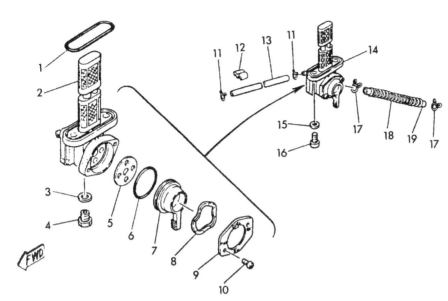

Fig. 7.7 Vacuum petrol tap -
XS650 SE UK model and XS650
E/SE onwards US models (Sec 6)

 1   O-ring
 2   Filter
 3   Sealing washer
 4   Drain bolt
 5   Tap valve
 6   O-ring
 7   Lever
 8   Wave washer
 9   Position plate
10   Screw - 2 off
11   Clip - 2 off
12   Pipe clamp
13   Vacuum pipe
14   Fuel tap
15   Washer - 2 off
16   Screw - 2 off
17   Clip - 2 off
18   Fuel pipe guard
19   Fuel pipe

8   Refit the remaining components, noting that the spacer must be free from any burrs which might damage the seal lips. Install a new lock washer, and tighten the nut to the specified torque setting; secure with the lock washer tab.

4   Clutch: modifications

All 1979-on UK models and XS650 F, 2F US models onwards

1   An improved method of securing the clutch nut to the mainshaft consists of a tabbed lock washer, the ears of which are bent up against the nut to lock it in position. The conical spring washer is still retained. Note that the lock washer should be renewed whenever the nut is disturbed.

XS650 SE UK models and XS650 G, H, SG, SH, SJ, SK US models

2   These models have a shock absorber assembly fitted to the clutch centre, and fewer clutch plates than fitted to previous models. The shock absorber consists of a spring seat, conical spring, inner plain plate and a wire retaining clip. Note that when working on the clutch, the order of these components should be the same as noted on removal and the inner plain plate must not be interchanged with any of the other plain plates.
3   Refer to Fig. 7.6 for details of the shock absorber assembly.

5   Gear selector mechanism: modification

1   Maintenance of the gearbox assembly remains identical to that given in the main text of this manual. It should be noted however, that the cam follower pin design has been changed for the later models to that of a single piece unit, thus discontinuing the use of the cam follower roller.

6   Vacuum petrol tap(s): description, removal and refitting - XS650 SE UK model and XS650 E/SE onwards US models

1   The vacuum tap houses a diaphragm which is opened by the vacuum created in the inlet manifold when the engine is running; a pipe connects the manifold and tap. In lever positions ON and RES the tap will only allow fuel to flow from the tank if the engine is running. The PRI position bypasses the diaphragm and allows fuel to flow freely irrespective of whether the engine is running or stopped.
2   The petrol filter is located inside the tank, on the tap stack. It follows that the tank must be drained and the tap withdrawn for access.
3   A drain bolt (or cover on early models) is situated in the tap base, its purpose being to allow sediment collected in the tap to be cleaned away without necessitating tap removal. With the tap in the ON or RES positions remove the cover and clean out the tap body; check the condition of the sealing washer and renew it if necessary.
4   To remove the tap(s), first drain the tank of fuel. With the lever in the ON or RES positions disconnect the fuel pipe from its union on the carburettors and insert the end of the pipe in a container suitable for the storage of petrol. Turn the tap lever to the PRI position and allow the fuel to drain.
5   With the tank fully drained of fuel, unscrew the 2 crosshead screws that retain the tap to the tank. In order to gain proper access to these screws it may be necessary to lift the seat and remove the fuel tank securing bolt so that the rear of the tank may be raised. Pull the tap away from the tank until the vacuum hose can be detached from the tap connection. The tap may now be fully withdrawn from the tank.
6   Servicing procedures for the tap assembly are as described in Chapter 2, Section 3, paragraphs 2, 3 and 4 of this manual. Reference should be made to Fig. 7.7 when carrying out these procedures.

7   Carburettors: modifications and fuel level check

Modifications

1   Carburettor changes throughout the model range of Yamaha's XS650 machines have been minimal, changes being made mainly for emission control purposes and to improve carburettor efficiency. The main changes made have been to alter the position of the main nozzle, pilot jet and pilot screw (see Fig. 7.8).
2   An internal air vent system was included on post 1978

Fig. 7.8 Carburettors (Sec 7)

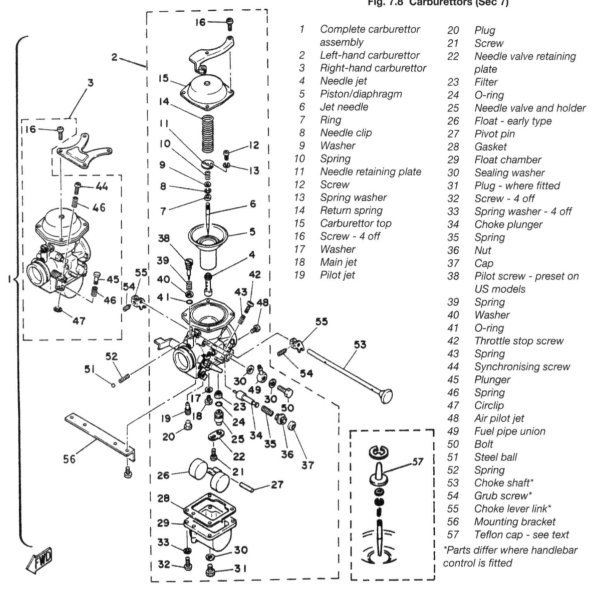

1	Complete carburettor assembly	20	Plug
2	Left-hand carburettor	21	Screw
3	Right-hand carburettor	22	Needle valve retaining plate
4	Needle jet	23	Filter
5	Piston/diaphragm	24	O-ring
6	Jet needle	25	Needle valve and holder
7	Ring	26	Float - early type
8	Needle clip	27	Pivot pin
9	Washer	28	Gasket
10	Spring	29	Float chamber
11	Needle retaining plate	30	Sealing washer
12	Screw	31	Plug - where fitted
13	Spring washer	32	Screw - 4 off
14	Return spring	33	Spring washer - 4 off
15	Carburettor top	34	Choke plunger
16	Screw - 4 off	35	Spring
17	Washer	36	Nut
18	Main jet	37	Cap
19	Pilot jet	38	Pilot screw - preset on US models
		39	Spring
		40	Washer
		41	O-ring
		42	Throttle stop screw
		43	Spring
		44	Synchronising screw
		45	Plunger
		46	Spring
		47	Circlip
		48	Air pilot jet
		49	Fuel pipe union
		50	Bolt
		51	Steel ball
		52	Spring
		53	Choke shaft*
		54	Grub screw*
		55	Choke lever link*
		56	Mounting bracket
		57	Teflon cap - see text

*Parts differ where handlebar control is fitted

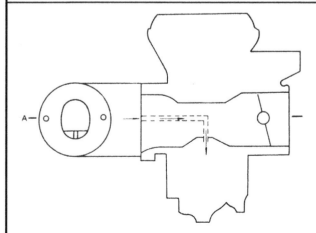

Fig. 7.9 Carburettor internal air vent system (Sec 7)

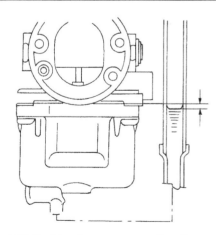

Fig. 7.10 Fuel level measurement (Sec 7)

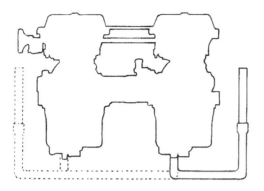

Fig. 7.11 Method of checking that machine is upright for fuel level check (Sec 7)

models, an air vent passage being provided between the float chamber and main bore. Also included was the introduction of a Teflon cap to cover the top end of the jet needle to stop the needle from vibrating during use.

3   The carburettor float assembly fitted to XS650 H, SH, SJ and SK US models is of different construction to the round brass float fitted to previous models. Additionally, the float chambers now have a drain screw fitted, which allows for the attachment of a fuel level checking device (see below for details).

4   On XS650 SJ and SK US models choke operation is by a lever on the left handlebar, via a cable, to the right-hand carburettor. The cable terminates in a trunnion type fitting at each end; it can be released from the handlebar lever end by removing the central screw from the underside of the lever. No cable adjuster is provided - if the cable has stretched to the point where full operation of the choke is not possible, the cable should be renewed. Apart from ensuring that the exposed parts of the inner cable are lubricated regularly, no maintenance is required.

Fuel level check

5   To measure the fuel level obtain either the calibrated Yamaha service tool or a length of transparent pipe with an internal bore of 6 mm (0.24 in). Position the motorcycle on its centre stand and connect the pipe end to the stub on the underside of the float chamber, the pipe should be held upright against the side

of the carburettor body (see Fig. 7.10). Tape can be used to hold the pipe in position, or attach a crocodile clip to the open end of the tube and part of the carburettor linkage.

6   Turn the fuel tap to ON or RES, start the engine and let it run for a minute to allow the level to stabilise. Open the drain screw to allow fuel to flow into the pipe, then measure the distance from the fuel level in the pipe to the gasket face of the float chamber - this should be 1 ± 1 mm (0.04 + 0.04 in). If outside this tolerance, the carburettors will need to be removed and the float height adjusted as described in Chapter 2.

7   Note that it is essential that the carburettors are exactly upright for this check, to ensure this, take the tube around to the outer side of the other carburettor and check the level (Fig. 7.11). If both are equal the machine is upright, if not, use pieces or card or plywood under the stand feet to level it.

8   Connect the tube to the other carburettor and check its level as described above.

8   Air filters: removal, cleaning and refitting - XS650 SE UK model and XS650 G, H, SG, SH, SJ, SK US models

1   To remove the foam/wire mesh filter element first unclip the outer side panel from its 3 retaining points. Remove the centre retaining screw from the inner side panel and detach the panel from the filter casing. Care should be taken to ensure the panel seal comes away cleanly with the panel and is not torn during removal.

2   The filter element may now be pulled out for cleaning. It should be noted that the forward seal on the element will provide some resistance to element removal if it has become adhered to the casing joint. Remove the element retaining spring along with the element (photo).

3   Clean the element as detailed in Chapter 2, Section 9 of this manual, depending on whether the dry or oiled type of foam element is fitted.

4   Refitting is a reversal of the removal procedure, noting the following points. The element retaining spring must be fitted correctly. The inner side panel seal must be correctly positioned in its retaining groove and be in good condition (photo).

9   Ignition timing: check - post 1978 models with contact breaker ignition

1   It should be noted when carrying out the ignition timing

8.2 The air filter forward seal may provide some resistance to element removal

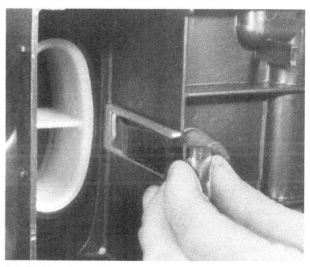

8.4 Position the retaining spring correctly before fitting the filter element

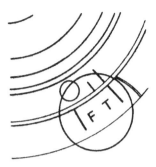

Fig. 7.12 Ignition timing marks - post 1978 models with contact breaker ignition (Sec 9)

sequence stated in Section 7, Chapter 3, that the alternator rotor and stator timing marks have been changed on post 1978 machines to those shown in Fig. 7.12. The F mark is no longer shown as an area between two marks but as a single mark.

10 Transistor Controlled Ignition (TCI): description - XS650 G, H, SG, SH, SJ and SK US models

1   This system comprises two main units; a pick-up unit fitted instead of the contact breaker assembly and an igniter unit.
2   Because the contact breaker assembly has been replaced by the pick-up unit which has no moving parts, frequent adjustment of the ignition timing is no longer necessary. The pick-up unit should not be disturbed unless renewal of the unit is required. Mechanical advance of the ignition is also eliminated by the incorporation of an automatic advance circuit built into the igniter unit.
3   The pick-up unit comprises two pick-up coils mounted on the generator stator and a permanent magnet attached to the crankshaft-mounted rotor. As the magnet passes through the field set up by the pick-up coils, signals are generated and passed to the igniter unit.
4   The igniter unit, located beneath the small cover under the left-hand side panel, provides the following system functions. A duty control circuit is provided to control the 'on time' period of the primary ignition current in order to reduce electrical consumption. A protective circuit for the ignition coil is also incorporated. Should the ignition switch be turned on and the crankshaft not turned, the protective circuit will stop current flowing to the primary coil within a few seconds. When the crankshaft is turned, the current will be turned on again by the signals generated by the pick-up coils. The unit advances the ignition timing electrically by using the two signals received from the pick-up coils.
Note: *On no account run the engine with the sparking plug cap(s) removed. Due to the high secondary voltage, it is possible to damage the internal insulation of the secondary coil.*

11 Transistor Controlled Ignition (TCI): testing - XS650 G, H, SG, SH, SJ and SK US models

1   Having no moving parts, the ignition system is likely to remain trouble-free during the life of the machine. It is, however, necessary to ensure that the battery is in good condition and fully charged, and that the spark plugs are clean and correctly gapped.
2   Trouble-shooting is straightforward if the fault is traced in a logical order. Readers will find the flow chart in Fig. 7.13 useful, especially if complete failure is experienced. If tracing the cause of a misfire which occurs only at high engine speed, it may be

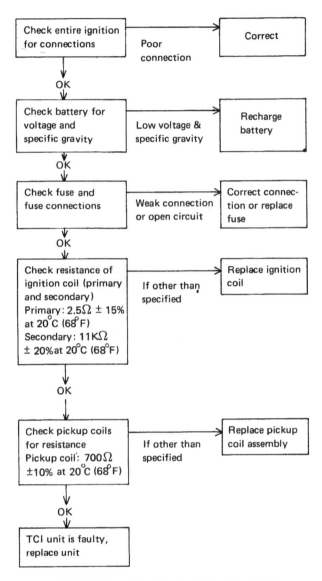

Fig. 7.13 Transistor Controlled Ignition (TCI) trouble-shooting flow chart (Sec 11)

advantageous to have the system tested on an electro tester by a Yamaha dealer. If the tests indicate a faulty pickup unit, note that on XS650 G and SG models the unit is secured by special shear-head screws and advice should be sought from a Yamaha dealer on their removal and the installation of a new pickup unit.
3   Ignition timing can be checked with a stroboscope using the marks on the generator rotor and stator, although it should be noted that the timing is not adjustable. A check of the timing, may however be a vital clue if the ignitor unit is suspected of malfunction.
4   To check the timing, first remove the generator inspection cover and identify the timing marks - there will be a single line on the rotor and an F mark (XS650 G and SG) or ⊔ mark (XS650 H, SH, SJ and SK) on the stator plate. Connect the stroboscope to No 1 cylinder following its maker's instructions, start the engine and run it at 1200 rpm whilst aiming the lamp at the timing marks. If the timing is correct, the line on the rotor should align with the F mark or be between the arms of the ⊔ mark (as applicable) on the stator plate.
5   The starter lockout circuit described in Sections 25 and 26 of this chapter is linked to the ignition circuit on later models. If a

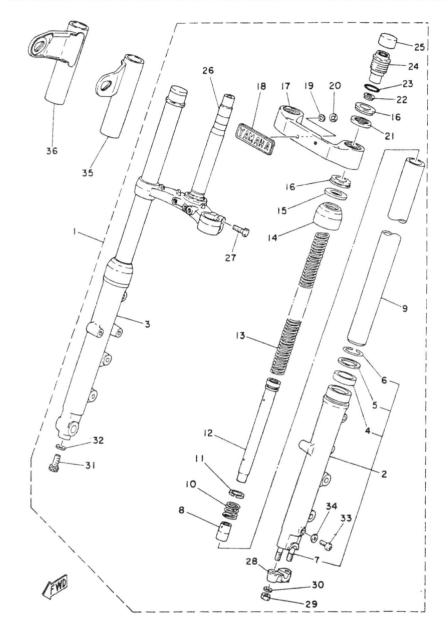

Fig. 7.14 Front forks - XS650 SE UK model and XS650 E/SE onwards US models (Sec 13)

1   Fork assembly
2   Lower leg assembly
3   Right-hand fork complete
4   Oil seal
5   Washer
6   Circlip
7   Stud - 2 off
8   Damper rod seat
9   Stanchion
10   Rebound spring
11   Sealing ring
12   Damper rod
13   Fork spring
14   Dust seal
15   Headlamp bracket washer
16   Headlamp bracket cushion - 2 off
17   Lower yoke cover*
18   Emblem
19   Washer - 2 off
20   Nut - 2 off
21   Headlamp bracket guide
22   Spring seat
23   O-ring
24   Cap bolt
25   Plastic cap
26   Lower yoke/steering stem
27   Pinch bolt
28   Wheel spindle clamp
29   Nut - 2 off
30   Washer - 2 off
31   Allen bolt
32   Sealing washer
33   Drain screw
34   Sealing washer
35   Left-hand headlamp bracket
36   Right-hand headlamp bracket
*Separate covers on certain models

fault is experienced which cannot be traced to any of the ignition components check the starter lockout circuit.

12 Steering head: modification - XS650 SJ and SK US models

1   If working on the steering head, note that the steering stem bolt is hidden under a cover carrying the 'Heritage Special' logo. Release the single screw from the rear of the cover and lift it free. The cover mounting bracket is secured to the steering stem bolt centre by a single screw.

13 Front forks: dismantling, reassembly and adjustment - XS650 SE UK models and XS650 E/SE onwards US models

Note: *Before removing the forks from the yokes, it is good practice to slacken the fork cap bolts whilst the legs are firmly held in the yokes.*

Dismantling

1   Work on one leg at a time to avoid the interchange of components. With the leg fully extended, fully unscrew the cap bolt whilst taking care to restrain it from being expelled forcibly under pressure from the spring as the last threads are freed. Lift out the spring seat and spring, then invert the leg to expel the oil. Release the dust seal from the top of the lower leg and slide this off the stanchion.

2   Remove the Allen bolt recessed in the bottom of the lower leg and pull the stanchion out of the lower leg. It may be found that once slackened, it is impossible to unscrew the Allen bolt due to rotation of the damper rod inside the stanchion and some means of holding the damper rod steady will have to be devised. Yamaha have a service tool available which consists of a slim rod with a shaped adaptor on its end to fit inside the recessed head of the damper rod. The tool is passed down through the stanchion, slotted into the damper rod and the other end held whilst the Allen bolt is slackened.

3   As an alternative to the service tool, try refitting the fork spring, spring seat and cap bolt and compressing the leg whilst the Allen bolt is unscrewed; the pressure of the spring may well

14.2 Release the dualseat rear spigots from their retaining slots in the frame - later Special models

hold the damper rod sufficiently for this purpose. If this fails, obtain a length of wooden dowel long enough to pass down through the stanchion and extend by at least 6 inches. Taper one end and press this into the damper rod head. Have an assistant hold the other end firmly and maintain downward pressure on the damper rod whilst you unscrew the Allen bolt.
4   On removal of the Allen bolt the stanchion can be withdrawn from the lower leg and the damper rod and rebound spring tipped out. The damper rod seat can be tipped out of the lower leg.
5   To renew the oil seal, remove the circlip and washer from the top of the lower leg. Using a large flat-bladed screwdriver prise the oil seal from position.
6   Refer to Chapter 4, Section 5 for details of component examination and renovation, noting the specifications given at the start of this chapter.

Reassembly

7   Before commencing reassembly ensure that all parts are clean and obtain new oil seals.
8   Lightly oil the outer edges of the new oil seal and press it squarely into the lower leg. If necessary use a piece of tubing, such as a socket, which bears only on the hard outer edge of the seal to drive it into position. Install the washer on top of the seal and fit the circlip into its groove.
9   Apply oil to the damper rod ring, slide the rebound spring over the rod and insert it into the stanchion. Fit the damper rod seat over the damper rod end which extends from the stanchion and slide the stanchion down into the lower leg; oil the stanchion surface to avoid damage to the oil seal lips. Clean the threads of the Allen bolt and ensure that its sealing washer is in place. Apply a few drops of thread-locking compound to its threads and screw the bolt into the damper rod end, using the method described previously to hold the damper rod steady. Tighten the bolt securely.
10 Slide the dust seal over the stanchion and fit it over the lower leg. Measure out the correct quantity and grade of fork oil and pour into the top of the stanchion. Fit the fork spring, spring seat and top cap. Tighten the top cap by hand, but it is advisable to wait until the leg is installed in the yokes before tightening it securely.

Adjustment

11 Spring preload adjustment allows the rider to select a setting to suit riding style and load carried. The 3-position adjuster is part of the cap bolt at the top of each fork leg.

12 To adjust, remove the plastic cap from the cap bolt head and pass a large screwdriver into the recessed adjuster. Rotate it clockwise to increase preload and anticlockwise to decrease it. *Always set each adjuster to the same position otherwise poor handling and loss of stability will result.*

14 Dualseat: removal and refitting

1   Dualseats fitted to the earlier Special models are retained to the frame of the machine by a similar method to that used on the Standard models. The removal and refitting procedures for this type of seat are therefore similar to those given in Chapter 4, Section 15 of this manual with the possible exception of a seat retaining rod assembly being fitted. This rod must be detached from one end-holder before seat removal is possible.
2   Dualseats fitted to the later Special models are fully detachable. Removal is achieved by unlocking the combined seat/helmet lock assembly mounted on the frame member to the rear right-hand side of the seat, then pulling the seat release lever fully rearwards to release the rear spigots from their retaining slots in the frame. The seat may then be pulled rearwards out of the front seat base spigot retaining slots and lifted up and away from the machine (photo).
3   It should be noted when refitting the seat that a sharp blow with the flat of the hand on the rear seat cushion may be needed to fully locate the rear spigots in their retaining slots.

15 Cast alloy wheels: examination and renovation

1   When examining cast alloy wheels, carefully check the complete wheel for cracks and chipping, particularly at the spoke roots and the edge of the rim. As a general rule a damaged wheel must be renewed as cracks will cause stress points which may lead to sudden failure under heavy load. Small nicks may be radiused carefully with a fine file and emery paper (No 600 - 1000) to relieve the stress. If there is any doubt as to the condition of a wheel, advice should be sought from a Yamaha repair specialist.
2   Each wheel is covered with a coating of lacquer to prevent corrosion. If damage occurs to the wheel and the lacquer finish is penetrated, the bared aluminium alloy will soon start to corrode. A whitish grey oxide will form over the damaged area, which in itself is a protective coating. This deposit however, should be removed carefully as soon as possible and a new protective coating of lacquer applied.
3   Check the lateral run-out at the rim by spinning the wheel and placing a fixed pointer close to the rim edge. If the maximum run-out is greater than 2.0 mm (0.08 in) Yamaha recommend that the wheel be renewed.
4   No means is available for straightening a warped wheel without resorting to the expense of having the wheel skimmed on all faces. If warpage was caused by impact during an accident, the safest measure is to renew the wheel complete. Worn wheel bearings may cause rim run-out. These should be renewed as described in Section 14 of Chapter 5.

16 Wheels: removal and refitting

1   Whether cast or spoked wheels are fitted, the bearing and spacer arrangement differs little between them, the method of rear braking being of more importance to the removal and refitting procedure. Reference to the wheel illustrations in this Chapter or those in Chapter 5 should be made in conjunction with the procedures described in Chapter 5.
2   On models with a rear disc brake, note that the caliper may be moved clear of the disc by slackening the torque arm bolt and lifting the caliper away from the disc after withdrawing the wheel spindle.

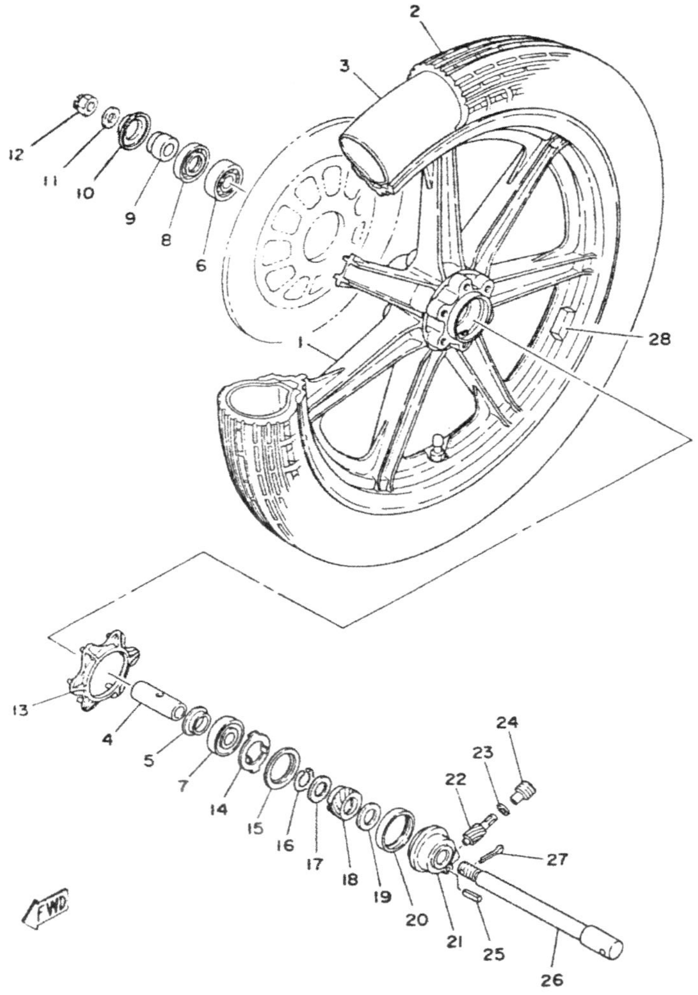

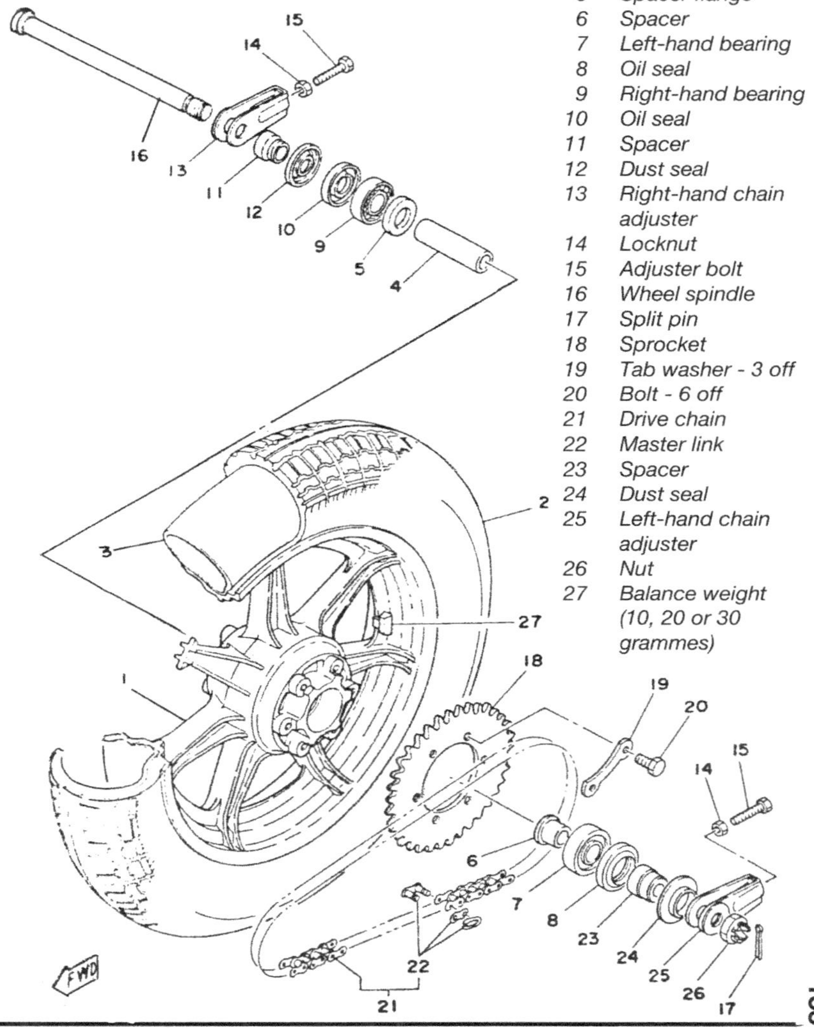

Fig. 7.16 Rear wheel - cast alloy type for disc brake (Sec 16)

1 Rear wheel
2 Tyre
3 Inner tube (where fitted)
4 Spacer
5 Spacer flange
6 Spacer
7 Left-hand bearing
8 Oil seal
9 Right-hand bearing
10 Oil seal
11 Spacer
12 Dust seal
13 Right-hand chain adjuster
14 Locknut
15 Adjuster bolt
16 Wheel spindle
17 Split pin
18 Sprocket
19 Tab washer - 3 off
20 Bolt - 6 off
21 Drive chain
22 Master link
23 Spacer
24 Dust seal
25 Left-hand chain adjuster
26 Nut
27 Balance weight (10, 20 or 30 grammes)

Fig. 7.15 Front wheel - cast alloy type (Sec 16)

1 Front wheel
2 Tyre
3 Inner tube (where fitted)
4 Spacer
5 Spacer flange
6 Right-hand wheel bearing
7 Left-hand wheel bearing
8 Oil seal
9 Spacer
10 Dust seal
11 Washer
12 Nut
13 Hub cover
14 Speedometer drive plate
15 Retainer
16 Circlip
17 Washer
18 Speedometer drive gear
19 Washer
20 Oil seal
21 Speedometer gear housing
22 Speedometer driven gear
23 Shim
24 Bush
25 Roll pin
26 Wheel spindle
27 Split pin
28 Balance weight (10, 20 or 30 grammes)

Fig. 7.17 Front brake master cylinder assembly (Sec 17)

1   Master cylinder
2   Primary piston assembly
3   Cover
4   Screw
5   Diaphragm
6   Plate
7   Handlebar clamp
8   Bolt
9   Spring washer
10 Brake lever
11 Adjuster bolt
12 Locknut
13 Spring
14 Lever pivot bolt
15 Nut
16 Cap
17 Union bolt
18 Plate washers
19 Brake hose
20 Hose boot
21 Front brake switch

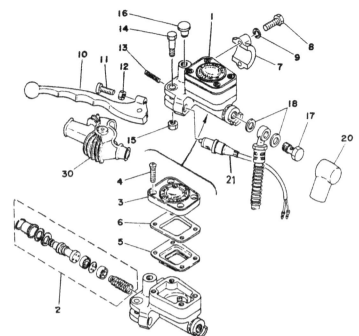

Fig. 7.18 Rear disc brake master cylinder (Sec 17)

1   Brake pedal
2   Pinch bolt
3   Spring washer
4   Spring
5   Brake pedal shaft
6   Pedal to rod link
7   Clevis pin
8   Washer
9   Split pin
10 Master cylinder rod
11 Locknut
12 Pedal height adjuster bolt
13 Locknut
14 Master cylinder
15 Primary piston assembly
16 Boot
17 Bolt
18 Spring washer
19 Fluid reservoir
20 Diaphragm
21 Plate
22 Cap
23 Bolt
24 Spring washer
25 Plain washer
26 Reservoir retainer
27 Mounting bracket
28 Spacer
29 Grommet
30 Plain washer
31 Screw
32 Spring washer
33 Reservoir to master cylinder hose
34 Hose guard
35 Clamp
36 Master cylinder to caliper hose
37 Union bolt
38 Plate washers

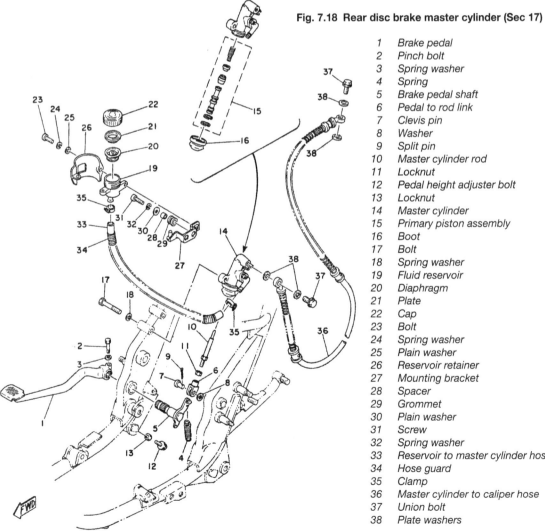

17 Braking system: modifications

Front brake

1 All later models have a single piston design front brake caliper and a square-shaped master cylinder. Both are shown in Figs. 7.17 and 7.19. Note that the standard fitting is for a single front disc brake, although a twin disc brake set up was available as an option on certain models.

Rear brake

2 A rear disc brake was fitted to all UK XS650 SE models and US XS650 SE, SF and SG models. The caliper is of the single piston type and is shown in Fig. 7.20. The master cylinder is located under the right-hand side panel and a check of its fluid level should be made at regular intervals as described for the front brake.

3 Later US Special models, reverted to a rear drum brake (see Sec 21 for details).

18 Front and rear brake calipers: pad renewal - single piston type

1 Brake pad renewal is greatly simplified on the single piston caliper. Pad wear inspection is also simplified by the inspection aperture situated in the rear of the caliper body (photo).

2 The pad renewal procedure is as follows. Remove the caliper support bolt and crossheaded pad securing screw and withdraw the caliper body from its frame. Lift the pads out of the mounting bracket, noting that on later models an anti-squeal shim will be fitted to the back of one or both pads. The anti-rattle spring may well become dislodged from the caliper when the pads are withdrawn.

3 Before fitting the new pads clean out their housings in the caliper body and mounting bracket, removing all traces of brake dust and dirt. *Although many new pads are asbestos free, take care not to inhale brake dust because it may be injurious to health - wear a face mask and perform the operation outside in a well-ventilated area.*

4 Apply copper-based grease sparingly to the pad edges and to the exposed lip of the caliper piston. Similarly apply a smear of this grease to the shank of the bolt which retains the caliper body to the bracket. Insert the new pads, together with anti-squeal shims (where fitted), into the mounting bracket, locating them with the springs top and bottom. Ensure the anti-rattle spring is in place, and refit the caliper body. Note the torque figures given in the Specifications at the beginning of this Chapter when reassembling the caliper unit.

19 Front and rear brake calipers: overhaul - single piston type

1 To dismantle the caliper first ensure it is drained of all fluid, then remove the caliper body from its mounting bracket by

18.1 Pad material can be inspected through aperture in rear of caliper

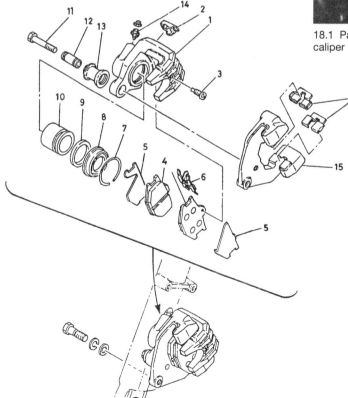

Fig. 7.19 Front brake caliper - single piston type (Sec 19)

1 Caliper body
2 Pad inspection cap
3 Fixed pad retaining screw
4 Pads
5 Anti-squeal shims (where fitted)
6 Anti-rattle spring
7 Circlip
8 Dust seal
9 Fluid seal
10 Piston
11 Caliper bolt
12 Sleeve
13 Bush
14 Bleed nipple
15 Mounting bracket
16 Pad springs

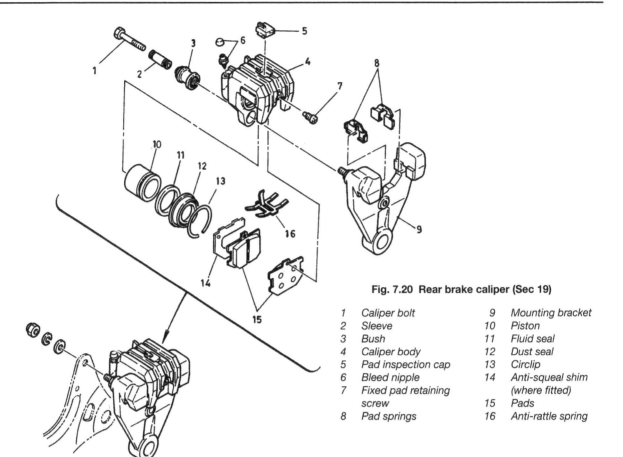

Fig. 7.20 Rear brake caliper (Sec 19)

1	Caliper bolt	9	Mounting bracket
2	Sleeve	10	Piston
3	Bush	11	Fluid seal
4	Caliper body	12	Dust seal
5	Pad inspection cap	13	Circlip
6	Bleed nipple	14	Anti-squeal shim
7	Fixed pad retaining		(where fitted)
	screw	15	Pads
8	Pad springs	16	Anti-rattle spring

undoing the retaining bolt and pad securing screw. Remove the piston retaining ring and dust seal and carefully displace the piston from the caliper with compressed air. **Warning:** *Before displacing the piston by this method, cover it with thick rag so that injury is avoided from the emerging piston and keep your fingers well away from the caliper aperture.* Never attempt to prise out the piston with pliers or similar. Remove the piston fluid seal.

2   Refer to Chapter 5 for details of caliper component examination and renovation.

3   When reassembling the caliper unit note the torque figures given in the Specifications at the beginning of this Chapter. If the pads have been disturbed, they should be refitted as described in the previous section. Take note of the lubrication advice given.

20 Rear disc brake: pedal height and free play adjustment

1   To adjust the rear brake pedal height, first loosen the adjuster bolt locknut situated on the frame bracket to the rear and slightly below the brake pedal pivot point. Turn the adjuster bolt clockwise or anti-clockwise so that the top of the pedal footplate is 12 - 18 mm (0.47 - 0.71 in) below the top of the footrest and tighten the locknut.

2   Loosen the master cylinder rod adjuster locknut and screw the rod downward until there is noticeable free play between the rod and master cylinder. Turn in the brake rod so that it lightly contacts the master cylinder then turn it out 1 - 1/5 turns to obtain the correct free play. This free play is measured at the forward end of the pedal and should be 13 - 15 mm (0.51 - 0.59 in) from when the pedal is moved to when the brake begins to take effect. Tighten the locknut and check that the punched mark on the master cylinder rod is not above the top surface of the locknut.

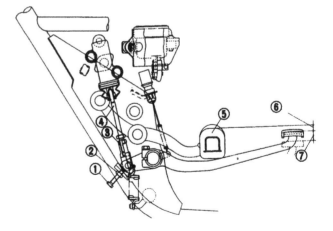

Fig. 7.21 Rear brake pedal height and free play adjustment - disc brake models (Sec 20)

1   Pedal height adjuster bolt
2   Adjuster bolt locknut
3   Master cylinder rod locknut
4   Master cylinder rod
5   Footrest
6   Pedal height measurement
7   Pedal free play measurement

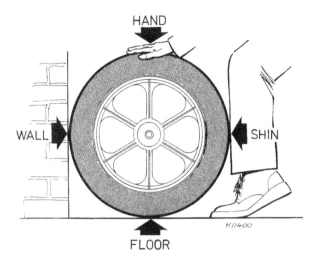

Fig. 7.22 Method of seating the beads on tubeless tyres (Sec 22)

---

21 Rear drum brake: general - 1981-on XS650 SE UK models and XS650 H, SH, SJ and SK US models

Wear check

1 Brake shoe lining wear can be assessed by the external wear indicator pointer - refer to Section 2 for details.

Adjustment

2 Prior to checking the pedal free play, check the pedal height, this being the distance from the top of the footrest to the top of the pedal. This can be set to suit personal preference anywhere within 12 - 18 mm (0.5 - 0.7 in). Adjustment is made via the bolt which bears on the pedal stop, after slackening its locknut. Tighten the locknut to secure.

3 Pedal free play should be checked as described in Chapter 5, Section 17.

Examination and renovation

4 Refer to Chapter 5, Section 17.

---

22 Tubeless tyres: general information, removal and refitting - XS650 SF, SG and SH US models

General information

1 These models may be fitted with tubeless tyres as original equipment. Such tyres will be marked 'TUBELESS' on the tyre sidewall, whereas the cast aluminum wheel will be marked 'SUITABLE FOR TUBELESS TYRES' on one of the wheel spokes. **Note:** *On no account must tubeless tyres be used on a wheel designed to take only tube-type tyres. Failure of the tyre and personal injury may result from sudden tyre deflation.*

2 A tubeless type wheel will take either a tubeless type of tyre or a tube-type tyre fitted with a tube. A tube type of wheel will only take a tube-type tyre. It is essential to ensure that when fitting a tube-type tyre, in both cases, that the correct type of tube is fitted.

3 It is strongly recommended that should a repair to a tubeless tyre be necessary, the wheel is removed from the machine and taken to a tyre fitting specialist or an authorized dealer. This is because the force required to break the seal between the wheel rim and tyre bead is considerable and likely to be beyond the capabilities of an individual working with normal tyre removing tools. Also, any abortive attempt to break the rim to bead seal

may cause damage to the wheel rim, resulting in expensive wheel renewal. If, however, a suitable bead releasing tool is available, and experience has already been gained in its use, tyre removal and refitting can be accomplished as follows.

Removal

4 Remove the wheel from the machine. Deflate the tyre by removing the valve core and when it is fully deflated, push the bead of the tyre away from the wheel rim on both sides so that the bead enters the well of the wheel rim. As noted, this operation will almost certainly require the use of a bead releasing tool.

5 Insert a tyre lever close to the valve and lever the edge of the tyre over the outside of the wheel rim. Very little force should be necessary; if resistance is encountered it is probably due to the fact that the tyre beads have not entered the well of the wheel rim all the way round the tyre. Should the initial problem persist, lubrication of the tyre bead and the inside edge and lip of the rim will facilitate removal. Use a recommended lubricant, a diluted solution of washing-up liquid (liquid detergent) or french chalk. Lubrication is usually recommended as an aid to tyre fitting but its use is equally desirable during removal. The risk of lever damage to wheel rims can be minimised by the use of proprietary plastic rim protectors placed over the rim flange at the point where the tyre levers are inserted. Suitable rim protectors can be fabricated very easily from short lengths (4 - 6 inches) of thick-walled nylon petrol pipe which have been split down one side using a sharp knife. The use of rim protectors should be adopted whenever levers are used and, therefore, when the risk of damage is likely.

6 Once the tyre has been edged over the wheel rim, it is easy to work around the wheel rim so that the tyre is completely free on one side.

7 Working from the other side of the wheel, ease the other edge of the tyre over the outside of the wheel rim, which is furthest away. Continue to work around the rim until the tyre is freed completely.

8 Refer to the following sections for details of puncture repair, tyre renewal and valves.

Refitting

9 Tyre refitting is virtually a reversal of the removal procedure. If the tyre has a balance mark (usually a spot of coloured paint), this must be positioned alongside the tyre valve. Similarly, any arrow indicating direction of rotation must face the right way.

10 Starting at the point furthest from the valve, push the tyre bead over the edge of the wheel rim until it is located in the well. Continue to work around the tyre in this fashion until the whole of one side of the tyre is on the rim. It may be necessary to use a tyre lever during the final stages. Here again, the use of a lubricant will aid fitting. It is strongly recommended that when fitting the tyre only a recommended lubricant is used because such lubricants also have sealing properties. Do not be over generous in the application of lubricant or tyre creep may occur.

11 Fitting the second bead is similar to fitting the first. Start by pushing the bead over the rim and into the well at a point diametrically opposite the tyre valve. Continue working around the tyre, each side of the starting point, ensuring that the bead opposite the working area is always in the well. Apply lubricant as necessary. Avoid using tyre levers unless absolutely essential, to help reduce damage to the soft wheel rim. Use of the levers should be required only when the final portion of bead is to be pushed over the rim.

12 Lubricate the tyre beads again prior to inflating the tyre, and check that the wheel rim is evenly positioned in relation to the tyre beads. Inflation of the tyre may well prove impossible without the use of a high pressure air hose. The tyre will retain air completely only when the beads are pressed firmly against the rim edges at all points and it may be found when using a foot pump that air escapes at the same rate as it is pumped in. This problem may also be encountered when using an air hose on new tyres which have been compressed in storage and by virtue of their profile hold the beads away from the rim edges. To

overcome this difficulty, a tourniquet may be placed around the circumference of the tyre, over the central area of the tread. The compression of the tread in this area will cause the beads to be pushed outwards in the desired direction. The type of tourniquet most widely used consists of a length of hose closed at both ends, with a suitable clamp fitted to enable both ends to be connected. An ordinary tyre valve is fitted at one end of the tube so that after the hose has been secured around the tyre it may be inflated, giving a constricting effect. Another possible method of seating beads to obtain initial inflation is to press the tyre into the angle between a wall and the floor. With the airline attached to the valve additional pressure is then applied to the tyre by the hand and shin, as shown in Fig. 7.22. The application of pressure at four points around the tyre's circumference whilst simultaneously applying the airline will often effect an initial seal between the tyre beads and wheel rim, thus allowing inflation to occur.

13 Having successfully accomplished inflation, increase the pressure to 40 psi and check that the tyre is evenly disposed on the wheel rim. This may be judged by checking that the thin positioning line found on each tyre wall is equidistant from the wheel rim around the total circumference of the tyre. If this is not the case, deflate the tyre, apply additional lubrication and reinflate. Minor adjustments to the tyre position may be made by bouncing the wheel on the ground.

14 Always run the tyres at the recommended pressures and never under- or over-inflate. The correct pressures are given in the Specifications at the start of this chapter. Note that if non-standard tyres are fitted check with the tyre manufacturer or supplier for recommended pressures. Finally refit the valve dust cap.

23 Tubeless tyres: puncture repair and tyre renewal - XS650 SF, SG and SH US models

1   If a puncture occurs, the tyre should be removed for inspection for damage before any attempt is made at remedial action. *The temporary repair of a punctured tyre by inserting a plug from the outside should not be attempted.* The manufacturers strongly recommend that no such repair is carried out on a motorcycle tyre. Not only does the tyre have a thin carcass, which does not give sufficient support to the plug, but the consequences of a sudden deflation are often sufficiently serious that the risk of such an occurrence should be avoided at all costs.

2   The tyre should be inspected both inside and out for damage to the carcass. Unfortunately the inner lining of the tyre - which takes the place of the inner tube - may easily obscure any damage and some experience is required in making a correct assessment of the tyre condition.

3   There are two main types of repair which are considered safe in repairing tubeless motorcycle tyres. The first consists of inserting a mushroom-headed plug into the hole from the inside of the tyre. The hole is prepared for insertion of the plug by reaming and the application of an adhesive. The second repair is carried out by buffing the inner lining in the damaged area and applying a cold or vulcanised patch. Because both inspection and repair, if they are to be carried out safely, require experience in this type of work, it is recommended that the tyre be placed in the hands of a repairer with the necessary skills, rather than repaired in the home workshop.

4   In the event of an emergency, the only recommended 'get-you-home' repair is to fit a standard inner tube of the correct size. If this course of action is adopted, care should be taken to ensure that the cause of the puncture has been removed before the inner tube is fitted. It may be found that the valve hole in the rim is considerably larger than the diameter of the inner tube valve stem. To prevent the ingress of road dirt, and to help support the valve, a spacer should be fitted over the valve.

5   In the event of the unavailability of tubeless tyres, ordinary

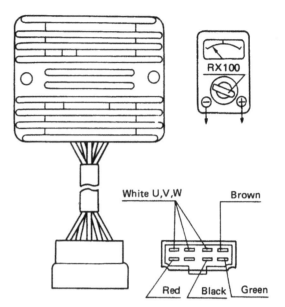

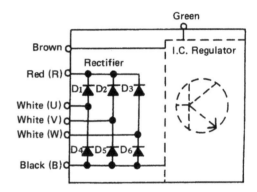

Checking element	Tester lead connecting point		Good	Replace (element shorted)	Replace (element opened)
	(+) (red)	(−) (black)			
D1	R	U	O	O	x
	U	R	x	O	x
D2	R	V	O	O	x
	V	R	x	O	x
D3	R	W	O	O	x
	W	R	x	O	x
D4	U	B	O	O	x
	B	U	x	O	x
D5	V	B	O	O	x
	B	V	x	O	x
D6	W	B	O	O	x
	B	W	x	O	x

Fig. 7.23 Rectifier test - XS650 G, H, SG, SH, SJ and SK US models (Sec 24)

O   *Continuity*   X   *No continuity (infinite resistance)*

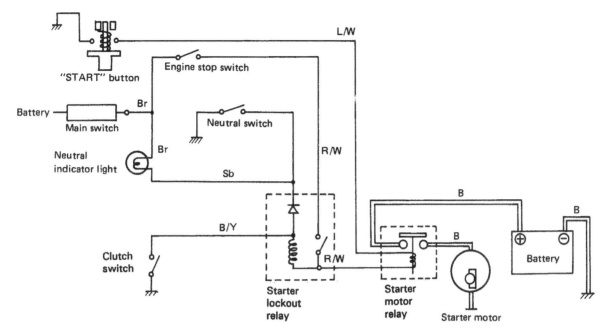

Fig. 7.24 Starter lockout circuit - XS650 H and SH US models (Sec 25)

See main wiring diagram for wire colour key

tubed tyres fitted with inner tubes of the correct size may be fitted. Refer to the manufacturer or a tyre fitting specialist to ensure that only a tyre and tube of equivalent type and suitability is fitted, and also to advise on the fitting of a valve nut/spacer to the rim hole.

---

24 Charging system: modifications - XS650 G, H, SG, SH, SJ and SK US models

Checking the charging system output

1   The battery must be known to be in good condition and fully charged for the test results to be accurate. This test measures the no-load output voltage, and therefore the lighting system should not be in operation whilst conducting the test; remove the headlamp circuit fuse to render the circuit inoperative. Connect a voltmeter set on the 0 - 20 volts dc scale across the battery, positive probe to positive terminal (+) and negative probe to negative terminal (-). Start the engine and run it at 2000 rpm or just above; the meter reading should be between 14.2 and 14.8 volts. Stop the engine as soon as the reading has been taken.
2   Adjustment of the regulator cannot be made on these models and any variation from the specified voltage range indicates failure of the regulator or an alternator fault. Check the alternator brushes, slip rings and windings before condemning the regulator.

Alternator testing

3   Refer to Chapter 6, Section 5 for details of the brush and slip ring condition checks.
4   To measure the winding resistance separate the multi-pin connector from the alternator and using an ohmmeter or multimeter set to the appropriate scale, measure the resistance across each pair of white wires (3 tests) on the alternator side of the connector; a reading of 0.46 ohm should be obtained in each case. If any test indicates infinite resistance or a reading well outside that specified the stator coil assembly is faulty.
5   The rotor coil winding can be checked in a similar manner, by making the meter connection across the green and brown wires

in the connector. A reading of 5.3 ohms should be shown; anything widely different indicates a faulty winding.

Voltage regulator/rectifier unit

6   An electronic regulator replaces the electro-mechanical unit fitted to previous models. It is a finned cast unit mounted on the left-hand side of the machine, below the lower side panel. Its performance can be checked by making the charging system output test described above.
7   The rectifier is integral with the regulator. To test, disconnect its wiring block connector and remove the unit from the machine. Using an ohmmeter or multimeter set on the ohms x 100 scale make the continuity tests shown in Fig. 7.23. If any diode shows continuity or no continuity in both directions then the complete unit must be renewed.

---

25 Starter lockout system: description - XS650 H and SH US models

1   The starter lockout circuit prevents operation of the starter motor unless the gearbox is in neutral, or if not, the clutch lever is pulled in. The starter circuit is shown in Fig. 7.24.
2   With the ignition switched on and the engine stop switch in the RUN position, power is fed to one terminal of the starter lockout relay. The contacts in this relay will remain off, preventing power to pass to the starter motor relay until neutral is selected, or the clutch lever is pulled in. Either condition will earth the windings of the starter lockout relay allowing its contacts to close and pass power through to the starter motor relay. If the starter button is then pressed, the starter motor will operate.
3   The starter lockout relay is located on the right side of the frame rear section (see Fig. 7.32). Remove the right-hand side panel for access. Note that its wire colours will serve as a guide to identification - see the wiring diagrams at the end of this chapter for details. No test details are available from the manufacturer, so if the unit is suspected of malfunction, and all other components in the circuit are satisfactory, it must be renewed.

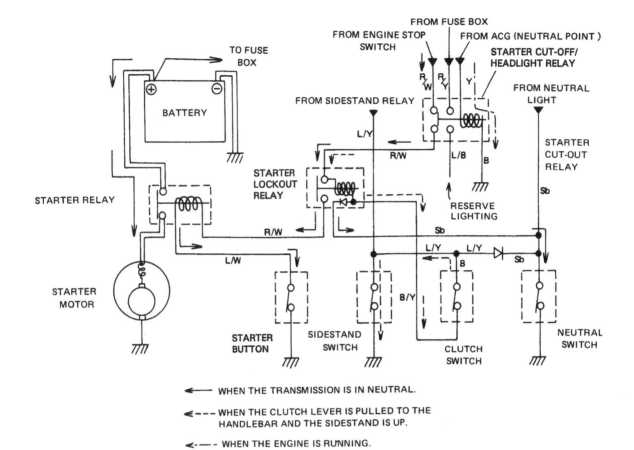

WHEN THE TRANSMISSION IS IN NEUTRAL.

WHEN THE CLUTCH LEVER IS PULLED TO THE
HANDLEBAR AND THE SIDESTAND IS UP.

WHEN THE ENGINE IS RUNNING.

7.25 Starter lockout circuit - XS650 SJ and SK US models (Sec 26)

See main wiring diagram for wire colour key

26 Starter lockout system: description and testing - XS650 SJ and SK US models

1   These models have a side stand switch incorporated in the circuit described in Section 25. Operation of the starter motor is only possible if the gearbox is in neutral, or if the clutch lever is pulled in and the side stand retracted (Fig. 7.25).

2   Due to its exposed position, most side stand switch faults will be as a result of the ingress of dirt or corrosion; the application of a water dispersant aerosol will usually cure this. If the switch is thought to have failed internally, trace its wires up to the block connector, then disconnect it and connector an ohmmeter across the blue/yellow and black wire terminals; continuity should be shown with the stand in the retracted position, and no continuity with it fully extended.

3   The side stand relay is located on the left-hand rear frame tube (Fig. 7.26). To test its winding, disconnect the block connector and measure the resistance across the terminals shown in Fig. 7.27. Operation of its contacts can be checked

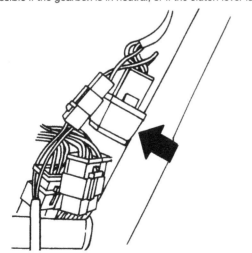

Fig. 7.26 Side stand relay location - XS650 SJ and SK US models (Sec 26)

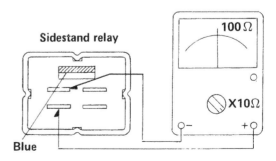

Fig. 7.27 Side stand relay winding check - XS650 SJ and SK US models (Sec 26)

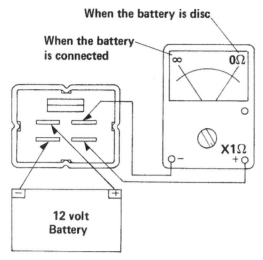

When the battery is disc

When the battery is connected

∞   0Ω

X1Ω

12 volt Battery

Fig. 7.28 Side stand relay contacts check - XS650 SJ and SK US models (Sec 26)

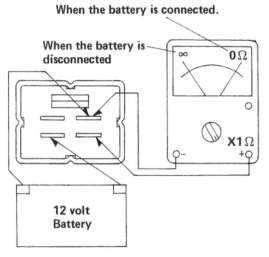

When the battery is connected.

When the battery is disconnected

∞   0Ω

X1Ω

12 volt Battery

Fig. 7.30 Starter lockout relay contacts check - XS650 SJ and SK US models (Sec 26)

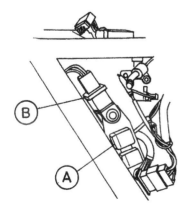

Fig. 7.32 Starter lockout circuit relay locations (Sec 26)

A   *Starter lockout relay*
B   *Starter cut-off/headlight relay (XS650 SJ and SK only)*

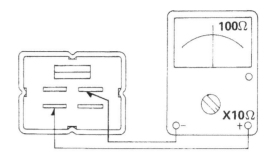

100Ω

X10Ω

Fig. 7.29 Starter lockout relay winding check - XS650 SJ and SK US models (Sec 26)

with a 12 volt battery connected as shown in Fig. 7.28.
4   Test details are available for the starter lockout relay fitted to this model. Its winding resistance, contacts and diode can be checked as shown in Figs. 7.29 to 7.31, although it should be noted that the value given for the diode is as obtained with a Yamaha test meter - other meters may produce slightly different readings.
5   An additional component in the circuit is the starter cut-off relay (the unit sharing its function with the headlight circuit) incorporated to protect the starter motor from operating whilst the engine is running. It receives power from the TCI unit when the engine is running, which earths through the cut-off relay, thus breaking the power to the starter circuit. The cut-off/headlight relay is located just above the starter lockout relay (Fig. 7.32). A test of its winding resistance and sets of contacts

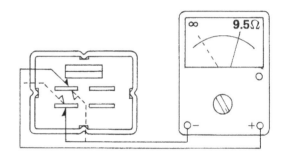

∞   9.5Ω

X1Ω

Fig. 7.31 Starter lockout relay diode check - XS650 SJ and SK US models (Sec 26)

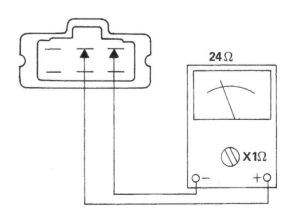

24Ω

X1Ω

Fig. 7.33 Starter cut-off/headlight relay winding check - XS650 SJ and SK US models (Sec 26)

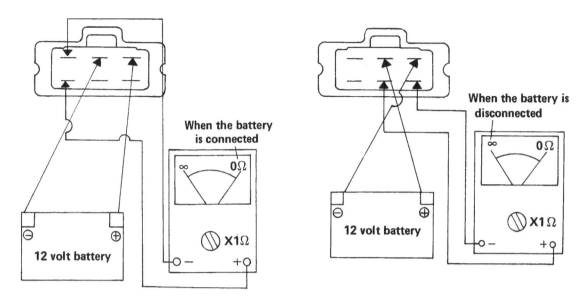

Fig. 7.34 Starter cut-off/headlight relay contacts check - XS650 SJ and SK US models (Sec 26)

can be made as shown in Figs. 7.33 and 7.34, although note that the value given for the winding is as obtained with a Yamaha test meter - other meters may produce slightly different readings.

27 Reserve lighting system: general - US models

1   This system has two functions, the first being to inform the rider (via a light in the instrument panel) that one of the headlamp filaments is inoperative, and the second being to automatically switch current to the remaining filament in the event of failure. Many models also feature a warning light for the stop lamp, and later models are fitted with a safety relay which automatically turns the lighting circuit on once the engine has been started.
2   The system varies in complexity depending on the model and reference to the wiring diagrams at the end of this chapter is advised if checking the circuit.

28 Self cancelling flashing indicators: description and testing

1   The purpose of this system is to turn off the turn signal automatically after a period of time or distance. At a very low speed the signal will cancel after a distance of 164 yards (150 metres) has been covered. At high speeds the signal will cancel after a time of 10 seconds has elapsed. When travelling in the lower range of speeds the signal will cancel after a combination of both time and distance.
2   If the handlebar switch lever is moved to the left or right turn positions it will return directly to the OFF position but the signal will continue to function until automatically cancelled electrically. By pushing the same lever in, the signal may be cancelled manually.
3   Should the system fail to operate, carry out the following test procedure. Pull off the 6-pin connector from the self cancelling unit and operate the handlebar switch. If the signals operate normally in the L, R and OFF positions then the bulbs, lighting circuit, handlebar switch light circuit and flasher unit are all in good operating condition.
4   If the previous check is satisfactory, then the flasher cancelling unit, the handlebar switch reset circuit or the

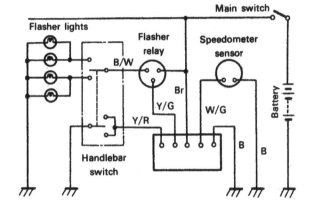

Fig. 7.35 Circuit diagram for self cancelling flashing indicators (Sec 28)

speedometer sensor circuit may be faulty. These components may be tested by carrying out the following test procedures.
5   Detach the 6-pin connector from the flasher cancelling unit and connect an ohmmeter with a 0 - 100 ohm range across the white/green and the black leads on the harness side. Turn the speedometer shaft. The ohmmeter needle should swing back and forth four times between zero and infinity on the scale, indicating that the speedometer sensor circuit is in good condition. If no needle deflection is apparent then the sensor or wire harness may be inoperative.
6   With the 6-pin connector detached from the flasher cancelling unit, check for continuity between the yellow/red lead on the harness side and the frame. With the handlebar switch set to the L or R position there should be a zero reading on the ohmmeter. With the switch set to the OFF position the needle should deflect to infinity. If the ohmmeter readings are not as stated, check the handlebar switch circuit and the wire harness for continuity.
7   If after completing the above checks the flasher cancelling system is still inoperative, then the cancelling unit must be renewed .
8   Should the flashing indicators operate only when the

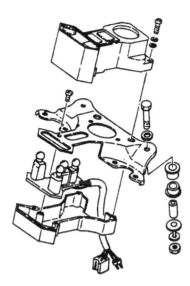

Fig. 7.36 Instrument panel - UK XS650 standard model and US XS650 D, E, F models (Sec 29)

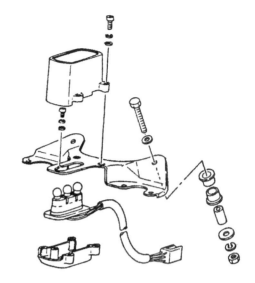

Fig. 7.37 Instrument panel - UK XS650 SE models and US XS650 2F, G, SG, H, SH, SJ, SK models (Sec 29)

handlebar switch lever is moved to the L or R positions and turn off immediately the switch lever returns to the centre position, renew the cancelling unit.

29 Instrument panel: modification

1 Whereas the speedometer and tachometer assemblies have remained identical to those of earlier models, the warning light console has been subject to various styling changes with an aim to improving warning light visibility.
2 Reference should be made to Fig. 7.36 or 7.37 (as applicable) when dismantling or reassembling the components, and to the wiring diagrams and specifications for bulb positions and ratings.

30 Stop and tail lamp: bulb renewal - 1980-on XS650 SE UK model and XS650 SG, H, SH, SJ, SK US models

1 The combined stop and tail lamp assembly is located in a rubber mounted holder mounted on the rear of the seat/grabrail unit and contains 2 bulbs, each bulb having 2 filaments, one for the stop lamp and one for the tail lamp.
2 The bulbs may be renewed after the plastic lens cover and 2 fixing screws have been removed.

31 Licence plate lamp: bulb renewal - 1980-on XS650 SE UK model and XS650 SG, H, SH, SJ, SK US models

1 Licence plate illumination is by a separate unit mounted on

30.2 Remove lens cover to gain access to stop and tail lamp bulbs -- later models

the rear mudguard. Two capless bulbs are fitted to US models and a single bayonet bulb on UK models.
2 To gain access to the bulb(s), remove the two nuts and pull the shield/lens assembly off its mounting. Either pull the bulbs (capless type) or twist (bayonet type) to remove them. When refitting, ensure the rubber damping washers are fitted correctly between the lamp and main mounting bracket.

Component key - XS650 standard UK model

1	Main switch	13	Safety relay	25	Horn	37	Meter light
2	RH handlebar switch	14	Light checker	26	Cancelling unit	38	Speedometer
3	LH handlebar switch	15	Rear indicators	27	Frame earth	39	Neutral light
4	Starter button	16	Fuse	28	Regulator	40	Brake light failure light
5	Engine stop switch	17	Battery	29	Rectifier	41	Indicator warning light RH
6	Headlamp switch	18	Starter button	30	Neutral switch	42	Indicator warning light LH
7	Dimmer switch	19	Starter motor	31	Generator	43	Pilot box
8	Horn button	20	Ignition coil	32	Front indicators	44	Meter light
9	Indicator switch	21	Tail/stop lamp	33	Headlamp	45	Meter light
10	Flash button	22	Condenser	34	Front brake switch	46	High beam light
11	Key removal possible	23	Contact breaker	35	Distance sensor	47	Tachometer
12	Rear brake switch	24	Flasher relay	36	Meter light		

Component key - XS650 SE UK model

1	Main switch	13	Fuses	25	Horn	37	Meter light
2	RH handlebar switch	14	Light checker	26	Cancelling unit	38	Speedometer
3	LH handlebar switch	15	Safety relay	27	Frame earth	39	Neutral light
4	Starter button	16	Rear indicators	28	Regulator	40	Brake/tail light failure light
5	Engine stop switch	17	Battery	29	Rectifier	41	Indicator warning light
6	Headlamp switch	18	Starter button	30	Neutral switch	42	Pilot box
7	Dimmer switch	19	Starter motor	31	Generator	43	High beam light
8	Horn button	20	Ignition coil	32	Front indicators	44	Meter light
9	Indicator switch	21	Tail/stop lamp	33	Headlamp	45	Meter light
10	Flash button	22	Condenser	34	Parking lamp	46	Tachometer
11	Key removal possible	23	Contact breaker	35	Distance sensor	47	Front brake switch
12	Rear brake switch	24	Flasher relay	36	Meter light		

Component key - XS650 D US model

1	Main switch	15	Meter light	29	Safety relay	43	Generator
2	RH handlebar switch	16	Meter light	30	Light checker	44	Neutral switch
3	LH handlebar switch	17	Pilot box	31	Reserve lighting unit	45	Rectifier
4	Starter button	18	Headlamp failure light	32	Resistor	46	Regulator
5	Engine stop switch	19	Indicator warning light RH	33	Brake show sensor	47	Frame earth
6	Headlamp switch	20	Indicator warning light LH	34	Diode	48	Cancelling unit
7	Dimmer switch	21	Stop lamp failure light	35	Rear indicators	49	Horn
8	Horn button	22	Neutral light	36	Fuse	50	Flasher unit
9	Indicator switch	23	Speedometer	37	Battery	51	Contact breaker
10	Key removal possible	24	Meter light	38	Starter button	52	Condenser
11	Key removal possible	25	Meter light	39	Starter motor	53	Tail lamp
12	Tachometer	26	Distance sensor	40	Ignition coil	54	Spark plugs
13	Brake shoe failure light	27	Front brake switch	41	Headlamp		
14	High beam light	28	Rear brake switch	42	Front indicators		

Component key - XS650 E and F US models

1	Main switch	14	Meter light	27	Rear brake switch	40	Neutral switch
2	RH handlebar switch	15	Meter light	28	Safety relay	41	Rectifier
3	LH handlebar switch	16	Pilot box	29	Light checker	42	Regulator
4	Starter button	17	Headlamp failure light	30	Reserve lighting unit	43	Frame earth
5	Engine stop switch	18	Indicator warning light RH	31	Rear indicators	44	Cancelling unit
6	Headlamp switch	19	Indicator warning light LH	32	Fuse	45	Horn
7	Dimmer switch	20	Stop lamp failure light	33	Battery	46	Flasher unit
8	Horn button	21	Neutral light	34	Starter button	47	Contact breaker
9	Indicator switch	22	Speedometer	35	Starter motor	48	Condenser
10	Key removal possible	23	Meter light	36	Ignition coil	49	Tail lamp
11	Key removal possible	24	Meter light	37	Headlamp	50	Spark plugs
12	Tachometer	25	Distance sensor	38	Front indicators		
13	High beam light	26	Front brake switch	39	Generator		

Colour key

B	Black	Lg	Light green
Br	Brown	O	Orange
Ch	Dark brown	P	Pink
Dg	Dark green	R	Red
G	Green	Sb	Light blue
Gy or Gr	Grey	W	White
L	Blue	Y	Yellow

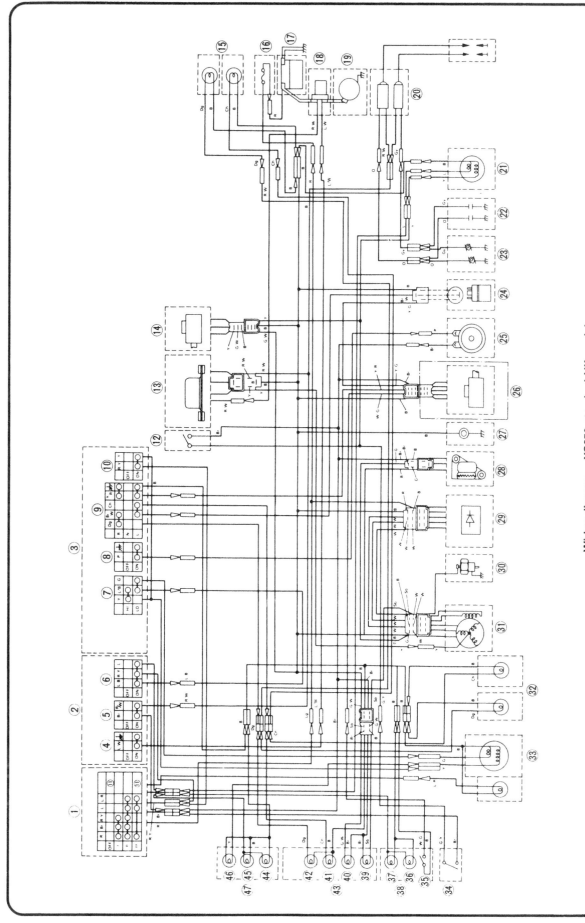

Wiring diagram – XS650 standard UK model
See page 164 for key

Wiring diagram – XS650 SE UK model
See page 164 for key

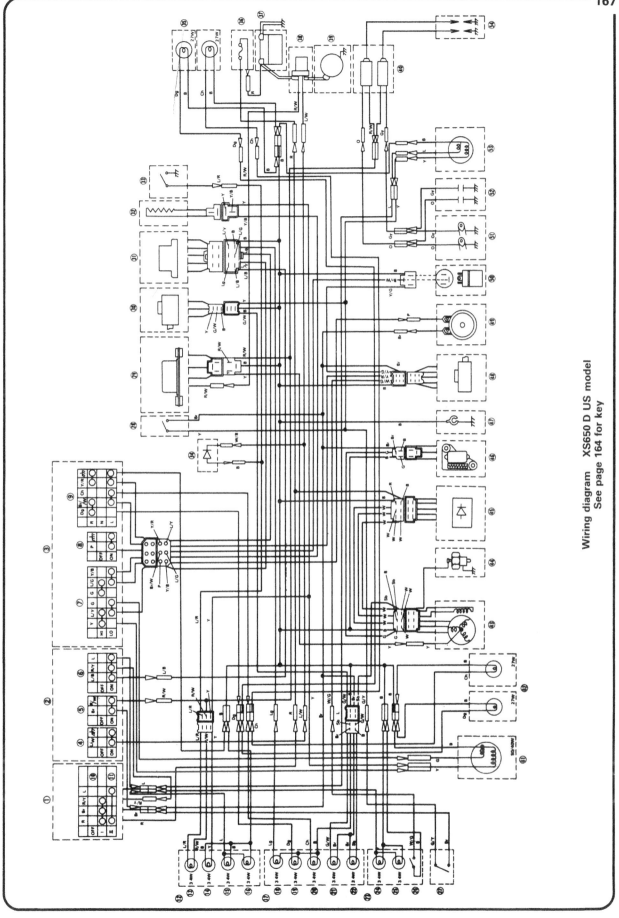

Wiring diagram XS650 D US model
See page 164 for key

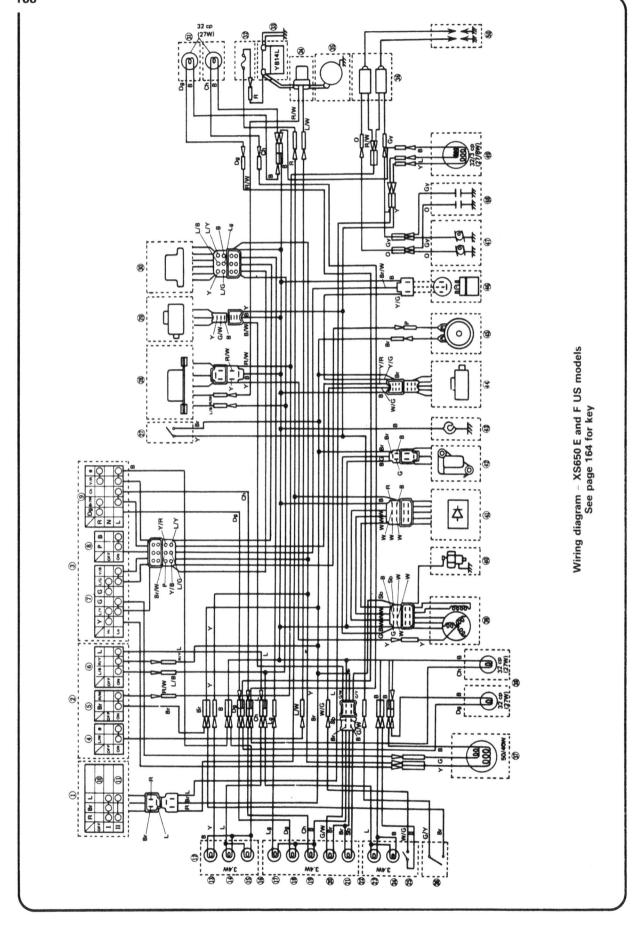

Wiring diagram – XS650 E and F US models
See page 164 for key

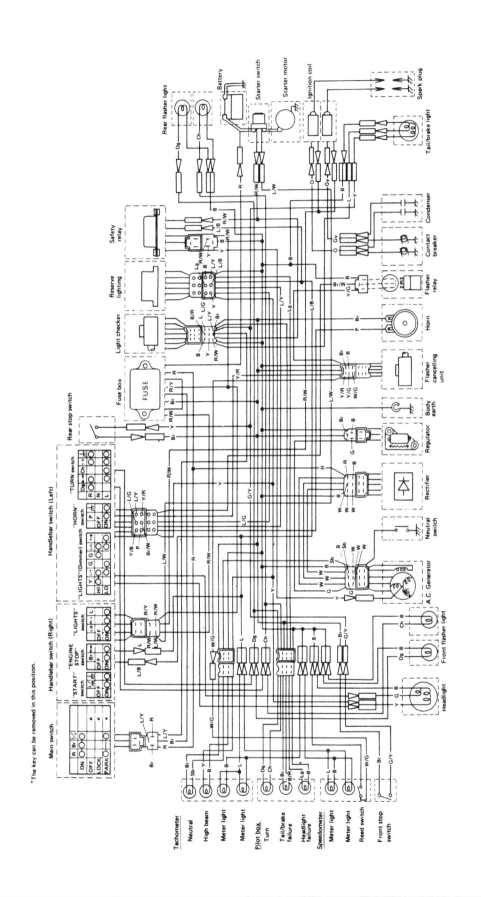

Wiring diagram – XS650 SE US model
See page 164 for key

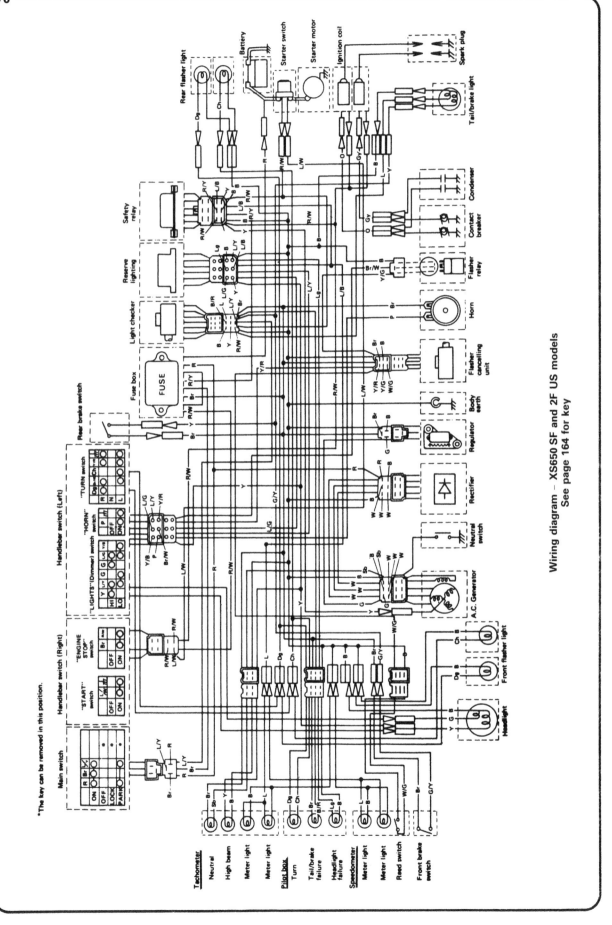

Wiring diagram – XS650 SF and 2F US models
See page 164 for key

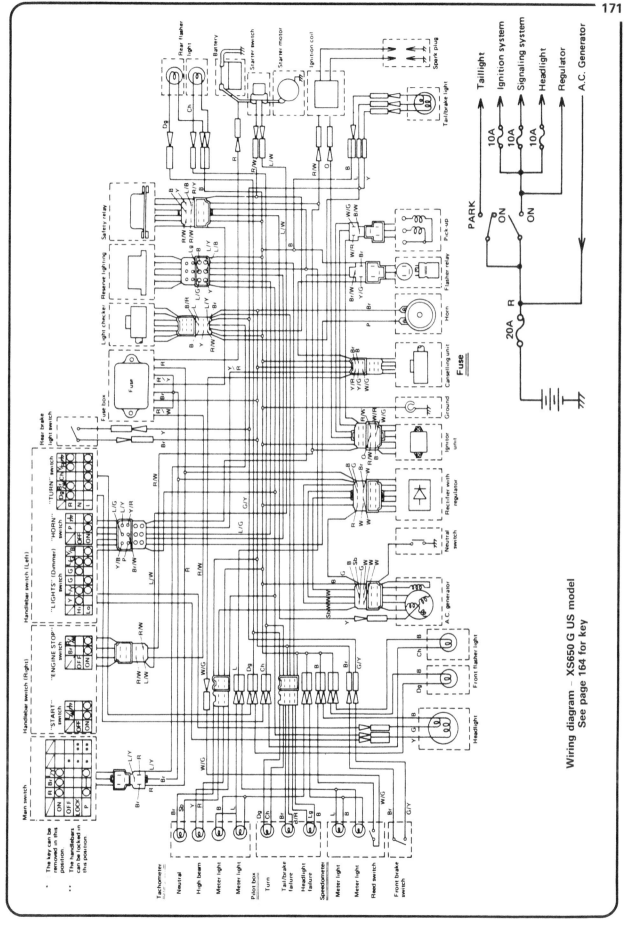

Wiring diagram – XS650 G US model
See page 164 for key

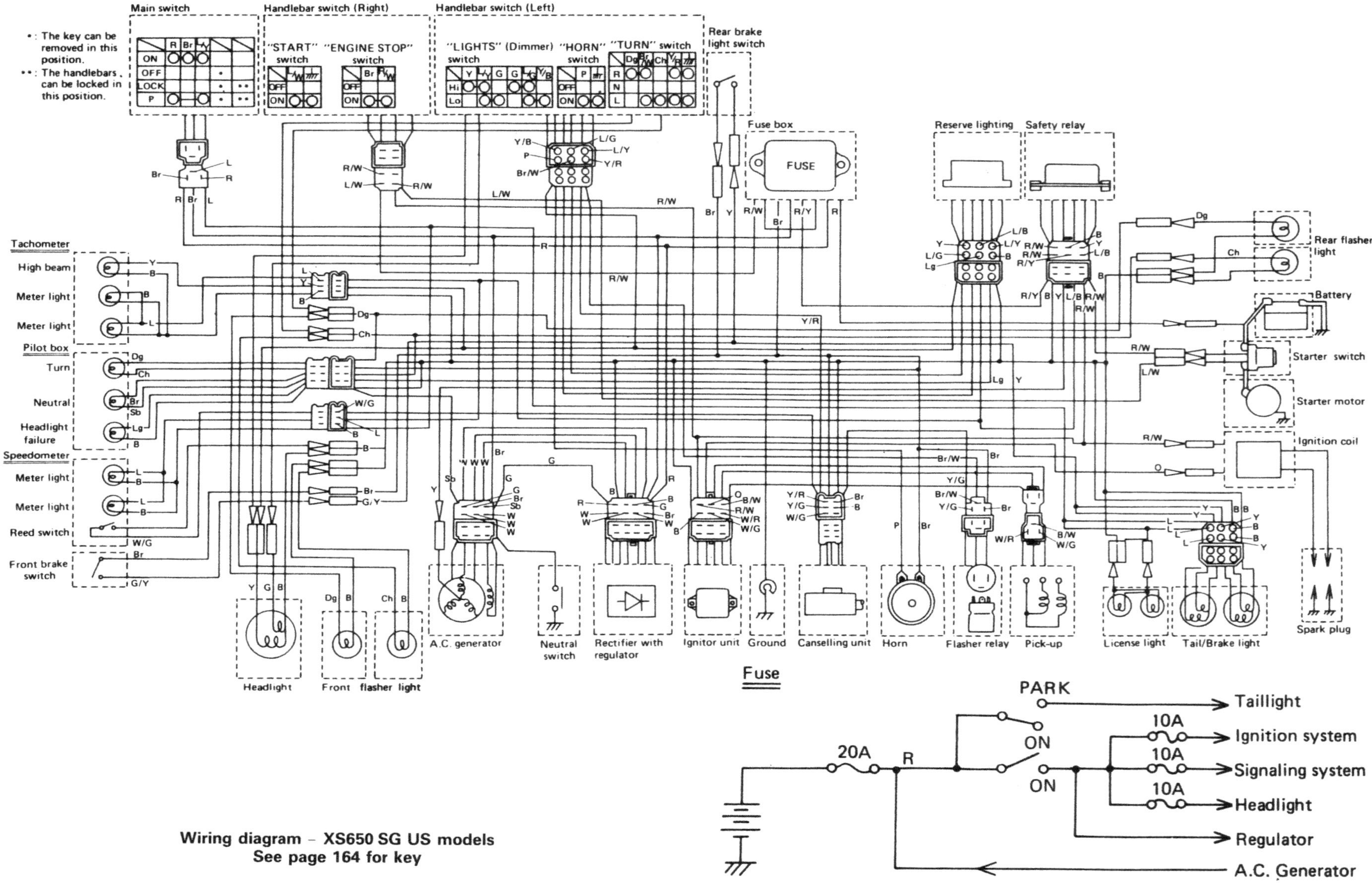

Rear flasher light

Battery

Starter switch

Starter motor

Ignition coil

Spark plug

Ch

Dg

Tail/Brake light

License light

Safety relay

Reserve lighting

Pick-up

Flasher relay

Horn

Canselling unit

Ground

Fuse box

FUSE

Ignitor unit

Rectifier with regulator

Rear brake light switch

Handlebar switch (Left)

"LIGHTS" (Dimmer) "HORN" "TURN" switch

	Dg		Y/B
R		N	
	P		L

"LIGHTS" (Dimmer) switch

Y	W	G	B
Hi			
Lo			

Neutral switch

A.C. generator

Handlebar switch (Right)

"START" "ENGINE STOP" switch

Br	F	W
OFF		
ON		

Main switch

	R	Br	W
ON			
OFF			
LOCK			
P			

· The key can be removed in this position.
·· The handlebars can be locked in this position.

Tachometer
High beam
Meter light
Meter light
Pilot box
Turn
Neutral
Headlight failure
Speedometer
Meter light
Meter light
Reed switch
Front brake switch

Front flasher light

Headlight

Wiring diagram – XS650 SG US models
See page 164 for key

PARK

Taillight
Ignition system
Signaling system
Headlight
Regulator
A.C. Generator

ON
ON

10A
10A
10A

R

20A

Fuse

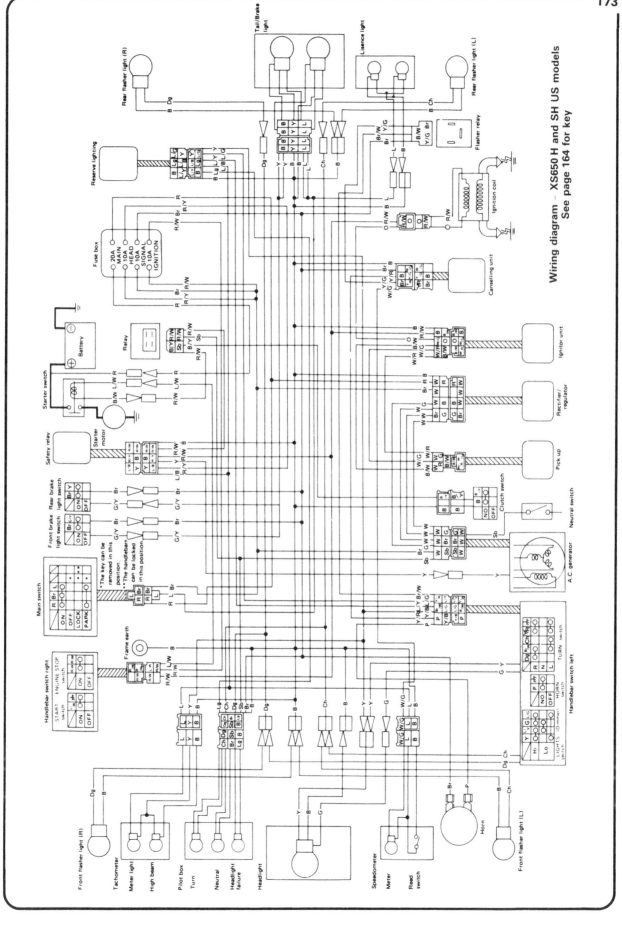

Wiring diagram – XS650 H and SH US models
See page 164 for key

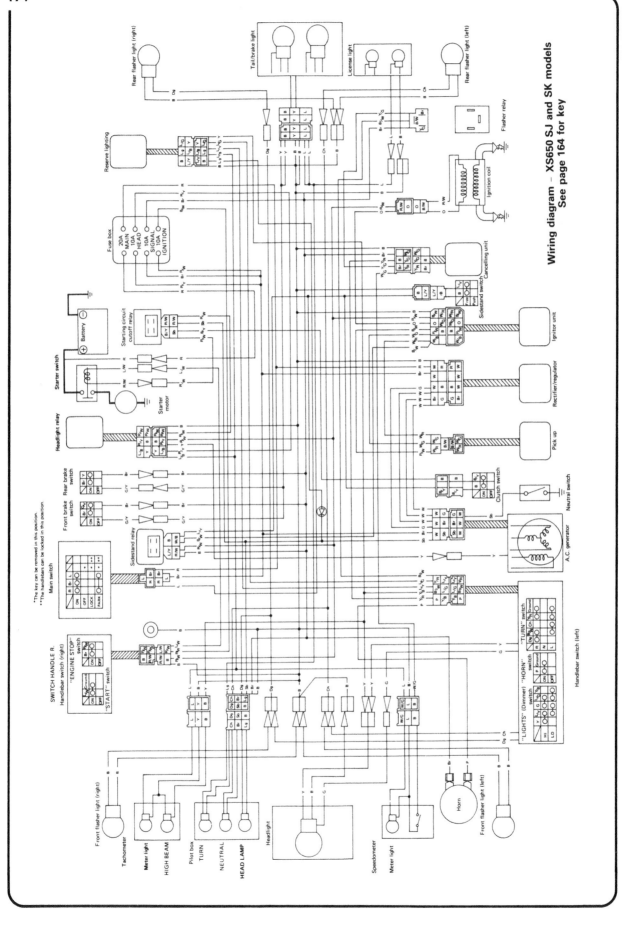

**Wiring diagram – XS650 SJ and SK models
See page 164 for key**

Conversion factors

Length (distance)
Inches (in)	X	25.4	= Millimetres (mm)	X	0.0394	= Inches (in)
Feet (ft)	X	0.305	= Metres (m)	X	3.281	= Feet (ft)
Miles	X	1.609	= Kilometres (km)	X	0.621	= Miles

Volume (capacity)
Cubic inches (cu in; in³)	X	16.387	= Cubic centimetres (cc; cm³)	X	0.061	= Cubic inches (cu in; in³)
Imperial pints (Imp pt)	X	0.568	= Litres (l)	X	1.76	= Imperial pints (Imp pt)
Imperial quarts (Imp qt)	X	1.137	= Litres (l)	X	0.88	= Imperial quarts (Imp qt)
Imperial quarts (Imp qt)	X	1.201	= US quarts (US qt)	X	0.833	= Imperial quarts (Imp qt)
US quarts (US qt)	X	0.946	= Litres (l)	X	1.057	= US quarts (US qt)
Imperial gallons (Imp gal)	X	4.546	= Litres (l)	X	0.22	= Imperial gallons (Imp gal)
Imperial gallons (Imp gal)	X	1.201	= US gallons (US gal)	X	0.833	= Imperial gallons (Imp gal)
US gallons (US gal)	X	3.785	= Litres (l)	X	0.264	= US gallons (US gal)

Mass (weight)
Ounces (oz)	X	28.35	= Grams (g)	X	0.035	= Ounces (oz)
Pounds (lb)	X	0.454	= Kilograms (kg)	X	2.205	= Pounds (lb)

Force
Ounces-force (ozf; oz)	X	0.278	= Newtons (N)	X	3.6	= Ounces-force (ozf; oz)
Pounds-force (lbf; lb)	X	4.448	= Newtons (N)	X	0.225	= Pounds-force (lbf; lb)
Newtons (N)	X	0.1	= Kilograms-force (kgf; kg)	X	9.81	= Newtons (N)

Pressure
Pounds-force per square inch (psi; lbf/in²; lb/in²)	X	0.070	= Kilograms-force per square centimetre (kgf/cm²; kg/cm²)	X	14.223	= Pounds-force per square inch (psi; lbf/in²; lb/in²)
Pounds-force per square inch (psi; lbf/in²; lb/in²)	X	0.068	= Atmospheres (atm)	X	14.696	= Pounds-force per square inch (psi; lbf/in²; lb/in²)
Pounds-force per square inch (psi; lbf/in²; lb/in²)	X	0.069	= Bars	X	14.5	= Pounds-force per square inch (psi; lbf/in²; lb/in²)
Pounds-force per square inch (psi; lbf/in²; lb/in²)	X	6.895	= Kilopascals (kPa)	X	0.145	= Pounds-force per square inch (psi; lbf/in²; lb/in²)
Kilopascals (kPa)	X	0.01	= Kilograms-force per square centimetre (kgf/cm²; kg/cm²)	X	98.1	= Kilopascals (kPa)
Millibar (mbar)	X	100	= Pascals (Pa)	X	0.01	= Millibar (mbar)
Millibar (mbar)	X	0.0145	= Pounds-force per square inch (psi; lbf/in²; lb/in²)	X	68.947	= Millibar (mbar)
Millibar (mbar)	X	0.75	= Millimetres of mercury (mmHg)	X	1.333	= Millibar (mbar)
Millibar (mbar)	X	0.401	= Inches of water (inH₂O)	X	2.491	= Millibar (mbar)
Millimetres of mercury (mmHg)	X	0.535	= Inches of water (inH₂O)	X	1.868	= Millimetres of mercury (mmHg)
Inches of water (inH₂O)	X	0.036	= Pounds-force per square inch (psi; lbf/in²; lb/in²)	X	27.68	= Inches of water (inH₂O)

Torque (moment of force)
Pounds-force inches (lbf in; lb in)	X	1.152	= Kilograms-force centimetre (kgf cm; kg cm)	X	0.868	= Pounds-force inches (lbf in; lb in)
Pounds-force inches (lbf in; lb in)	X	0.113	= Newton metres (Nm)	X	8.85	= Pounds-force inches (lbf in; lb in)
Pounds-force inches (lbf in; lb in)	X	0.083	= Pounds-force feet (lbf ft; lb ft)	X	12	= Pounds-force inches (lbf in; lb in)
Pounds-force feet (lbf ft; lb ft)	X	0.138	= Kilograms-force metres (kgf m; kg m)	X	7.233	= Pounds-force feet (lbf ft; lb ft)
Pounds-force feet (lbf ft; lb ft)	X	1.356	= Newton metres (Nm)	X	0.738	= Pounds-force feet (lbf ft; lb ft)
Newton metres (Nm)	X	0.102	= Kilograms-force metres (kgf m; kg m)	X	9.804	= Newton metres (Nm)

Power
Horsepower (hp)	X	745.7	= Watts (W)	X	0.0013	= Horsepower (hp)

Velocity (speed)
Miles per hour (miles/hr; mph)	X	1.609	= Kilometres per hour (km/hr; kph)	X	0.621	= Miles per hour (miles/hr; mph)

Fuel consumption*
Miles per gallon, Imperial (mpg)	X	0.354	= Kilometres per litre (km/l)	X	2.825	= Miles per gallon, Imperial (mpg)
Miles per gallon, US (mpg)	X	0.425	= Kilometres per litre (km/l)	X	2.352	= Miles per gallon, US (mpg)

Temperature

Degrees Fahrenheit = (°C x 1.8) + 32   Degrees Celsius (Degrees Centigrade; °C) = (°F - 32) x 0.56

*It is common practice to convert from miles per gallon (mpg) to litres/100 kilometres (l/100km), where mpg (Imperial) x l/100 km = 282 and mpg (US) x l/100 km = 235

English/American terminology

Because this book has been written in England, British English component names, phrases and spellings have been used throughout. American English usage is quite often different and whereas normally no confusion should occur, a list of equivalent terminology is given below.

English	American	English	American
Air filter	Air cleaner	Number plate	License plate
Alignment (headlamp)	Aim	Output or layshaft	Countershaft
Allen screw/key	Socket screw/wrench	Panniers	Side cases
Anticlockwise	Counterclockwise	Paraffin	Kerosene
Bottom/top gear	Low/high gear	Petrol	Gasoline
Bottom/top yoke	Bottom/top triple clamp	Petrol/fuel tank	Gas tank
Bush	Bushing	Pinking	Pinging
Carburettor	Carburetor	Rear suspension unit	Rear shock absorber
Catch	Latch	Rocker cover	Valve cover
Circlip	Snap ring	Selector	Shifter
Clutch drum	Clutch housing	Self-locking pliers	Vise-grips
Dip switch	Dimmer switch	Side or parking lamp	Parking or auxiliary light
Disulphide	Disulfide	Side or prop stand	Kick stand
Dynamo	DC generator	Silencer	Muffler
Earth	Ground	Spanner	Wrench
End float	End play	Split pin	Cotter pin
Engineer's blue	Machinist's dye	Stanchion	Tube
Exhaust pipe	Header	Sulphuric	Sulfuric
Fault diagnosis	Trouble shooting	Sump	Oil pan
Float chamber	Float bowl	Swinging arm	Swingarm
Footrest	Footpeg	Tab washer	Lock washer
Fuel/petrol tap	Petcock	Top box	Trunk
Gaiter	Boot	Torch	Flashlight
Gearbox	Transmission	Two/four stroke	Two/four cycle
Gearchange	Shift	Tyre	Tire
Gudgeon pin	Wrist/piston pin	Valve collar	Valve retainer
Indicator	Turn signal	Valve collets	Valve cotters
Inlet	Intake	Vice	Vise
Input shaft or mainshaft	Mainshaft	Wheel spindle	Axle
Kickstart	Kickstarter	White spirit	Stoddard solvent
Lower leg	Slider	Windscreen	Windshield
Mudguard	Fender		

Index